Introductory Mycology

Introductory Mycology

Dorian Snyder

Larsen & Keller
www.larsen-keller.com

Introductory Mycology
Dorian Snyder
ISBN: 978-1-64172-080-9 (Hardback)

⊟ Larsen & Keller

Published by Larsen and Keller Education,
5 Penn Plaza,
19th Floor,
New York, NY 10001, USA

Cataloging-in-Publication Data

Introductory mycology / Dorian Snyder.
    p. cm.
Includes bibliographical references and index.
ISBN 978-1-64172-080-9
1. Mycology. 2. Botany. I. Snyder, Dorian.
QK603 .I58 2019
579.5--dc23

For more information regarding Larsen and Keller Education and its products, please visit the publisher's website www.larsen-keller.com

Table of Contents

Permissions

Index

Preface

Mycology is the branch of biology that is concerned with the study of fungi, their genetic and biochemical properties, their taxonomy and applications for human use. Fungi can be both harmful and beneficial to humans. Fungi produce antibiotics, toxins and secondary metabolites. They can cause toxicity or infection, but they can also be a source of tinder, food, medicine and entheogens. Many species of mushrooms are cultivated for food, such as button mushrooms, Portobello mushrooms, oyster mushrooms and shiitakes, besides many others. Penicillin, lovastatin, griseofulvin, etc. are some drugs produced using fungi. Many varieties of fungi are used for the industrial production of vitamins, antibiotics and cholesterol-lowering drugs. Fungi can also be useful in suppressing plant pathogens like weeds, insects, mites, etc. in agriculture. This book provides comprehensive insights into the field of mycology. It aims to shed light on some of the unexplored aspects of this field. This textbook is appropriate for those seeking detailed information in this area.

A short introduction to every chapter is written below to provide an overview of the content of the book:

Chapter 1 - Mycology is the study of fungi, their properties and use for human benefit. Some of the sub-disciplines of mycology are mycotoxicology, ethnomycology and lichenology. The aim of this chapter is to explore mycology and its allied fields; **Chapter 2** - Fungi are eukaryotic heterotrophic organisms that acquire food by absorbing dissolved molecules. Some examples of fungi are mushrooms, molds and yeasts. The major groups of fungi are Chytridiomycota, Blastocladiomycota, Zygomycota, Ascomycota, Glomeromycota and Basidiomycota, which have been discussed in detail in this chapter; **Chapter 3** - Fungi generally grow as hyphae. These are thread-like cylindrical structures. This chapter contains detailed description of various fungal structures, such as hypha, sporocarp, conidium, crozier and veil. The topics elaborated in this chapter will help in gaining a better perspective about the anatomy and morphology of fungi; **Chapter 4** - Mycorrhiza refers to a symbiotic association between the roots of a vascular host plant and a fungus. In such an association, a fungus colonizes a host plant's root tissues. The aim of this chapter is to explore the different types of mycorrhizal associations, such as Arbuscular mycorrhiza, Ectomycorrhiza, Ericoid mycorrhiza, Orchid mycorrhiza, etc.; **Chapter 5** - Fungiculture refers to the cultivation of fungi for the production of food, medicine and other products. Many fungi are cultivated commercially across the globe, such as button mushroom, blewit, jelly fungi, winter mushroom, beech mushroom, shiitake, etc. This chapter has been carefully written to provide an in-depth understanding of fungiculture.

I extend my sincere thanks to the publisher for considering me worthy of this task. Finally, I thank my family for being a source of support and help.

Dorian Snyder

An Introduction to Mycology

Mycology is the study of fungi, their properties and use for human benefit. Some of the sub-disciplines of mycology are mycotoxicology, ethnomycology and lichenology. The aim of this chapter is to explore mycology and its allied fields.

Mycology is the study of fungi. It is closely associated with plant pathology as fungi cause the majority of plant disease.

Reasons for Mycology being Important

Fungi are the primary decomposers of organic material in many ecosystems and so play a crucial part in recycling nutrients and the global carbon cycle. They break down pollutants and the most durable organic materials and have a range of uses such as in medicine and food production. At least 80% of plants rely on mycorrhizal associations – symbiotic relationships between the plant's roots and a fungus that provides the plant with water and nutrients.

Fungi exist throughout the environment. Some of them are useful, for example as food or as the basis of medication. Others are less desirable, such as mold on food, or spores that cause diseases.

Neither plants nor animals, fungi belong to a group of their own. There are about 99,000 known species of fungal organisms, including yeasts, rusts, smuts, mildews, molds, and mushrooms.

Fungi are found in almost any habitat, including the International Space Station (ISS), where they were found to decompose food, with some spores surviving 5 months in microgravity.

Many live on the land, mainly in soil or on plant material. They are one of the most widely distributed organisms on the Earth.

They feature in foods, such as mushrooms and baker's yeast, and they have important roles in medicine and the environment.

Fungus

A fungus (plural: fungi or funguses) is any member of the group of eukaryotic organisms that includes microorganisms such as yeasts and molds, as well as the more familiar mushrooms. These organisms are classified as a kingdom, Fungi, which is separate from the other eukaryotic life kingdoms of plants and animals.

A characteristic that places fungi in a different kingdom from plants, bacteria, and some protists is chitin in their cell walls. Similar to animals, fungi are heterotrophs; they acquire their food by absorbing dissolved molecules, typically by secreting digestive enzymes into their environment. Fungi do not photosynthesise. Growth is their means of mobility, except for spores (a few of which are flagellated), which may travel through the air or water. Fungi are the principal decomposers in ecological systems. These and other differences place fungi in a single group of related organisms, named the *Eumycota* (*true fungi* or *Eumycetes*), which share a common ancestor (form a *monophyletic group*), an interpretation that is also strongly supported by molecular phylogenetics. This fungal group is distinct from the structurally similar myxomycetes (slime molds) and oomycetes (water molds). The discipline of biology devoted to the study of fungi is known as mycology. In the past, mycology was regarded as a branch of botany, although it is now known fungi are genetically more closely related to animals than to plants.

Abundant worldwide, most fungi are inconspicuous because of the small size of their structures, and their cryptic lifestyles in soil or on dead matter. Fungi include symbionts of plants, animals, or other fungi and also parasites. They may become noticeable when fruiting, either as mushrooms or as molds. Fungi perform an essential role in the decomposition of organic matter and have fundamental roles in nutrient cycling and exchange in the environment. They have long been used as a direct source of human food, in the form of mushrooms and truffles; as a leavening agent for bread; and in the fermentation of various food products, such as wine, beer, and soy sauce. Since the 1940s, fungi have been used for the production of antibiotics, and, more recently, various enzymes produced by fungi are used industrially and in detergents. Fungi are also used as biological pesticides to control weeds, plant diseases and insect pests. Many species produce bioactive compounds called mycotoxins, such as alkaloids and polyketides, that are toxic to animals including humans. The fruiting structures of a few species contain psychotropic compounds and are consumed recreationally or in traditional spiritual ceremonies. Fungi can break down manufactured materials and buildings, and become significant pathogens of humans and other animals. Losses of crops due to fungal diseases (e.g., rice blast disease) or food spoilage can have a large impact on human food supplies and local economies.

The fungus kingdom encompasses an enormous diversity of taxa with varied ecologies, life cycle strategies, and morphologies ranging from unicellular aquatic chytrids to large mushrooms. However, little is known of the true biodiversity of Kingdom Fungi, which has been estimated at 2.2 million to 3.8 million species. Of these, only about 120,000 have been described, with over 8,000 species known to be detrimental to plants and at least 300 that can be pathogenic to humans. Ever since the pioneering 18th and 19th century taxonomical works of Carl Linnaeus, Christian Hendrik Persoon, and Elias Magnus Fries, fungi have been classified according to their morphology (e.g., characteristics such as spore color or microscopic features) or physiology. Advances in molecular genetics have opened the way for DNA analysis to be incorporated into taxonomy, which has sometimes challenged the historical groupings based on morphology and other traits. Phylogenetic studies published in the last decade have helped reshape the classification within Kingdom Fungi, which is divided into one subkingdom, seven phyla, and ten subphyla.

Fungal Behavior

Fungi are capable of a variety of behaviors. Nearly all secrete chemicals, and some of these chemicals act as pheromones to communicate with other individuals. Many of the most dramatic examples involve mechanisms to get fungal spores dispersed to new environments. In mushrooms, spores are propelled into the air space between the gills, where they are free to fall down and can then be carried by air currents. Other fungi shoot spores aimed at openings in their surroundings, sometimes reaching distances over a meter.

Fungi such as Phycomyces blakesleeanus employ a variety of sensory mechanisms to avoid obstacles as their fruiting body grows, growing against gravity, toward light (even on the darkest night), into wind, and away from physical obstacles (probably using a mechanism of chemical sounding).

Other fungi form constricting rings or adhesive knobs that trap nematodes, which the fungus then digests.

One hormone that is used by many fungi is Cyclic adenosine monophosphate (cAMP).

Mycotoxicology

Mycotoxicology is the branch of mycology that focuses on analyzing and studying the toxins produced by fungi, known as mycotoxins. In the food industry it is important to adopt measures that keep mycotoxin levels as low as practicable, especially those that are heat-stable. These chemical compounds are the result of secondary metabolism initiated in response to specific developmental or environmental signals. This includes biological stress from the environment, such as lower nutrients or competition for those available. Under this secondary path the fungus produces a wide array of compounds in order to gain some level of advantage, such as incrementing the efficiency of metabolic processes to gain more energy from less food, or attacking other microorganisms and being able to use their remains as a food source.

Mycotoxins

Mycotoxins are toxic secondary metabolic products of molds present on almost all agricultural commodities worldwide. Unlike primary metabolites (sugars, amino acids and other substances), secondary metabolites are not essential in the normal metabolic function of the fungus. Other known secondary metabolites are phytotoxins and antibiotics.

Currently there are around 400 mycotoxins reported. These compounds occur under natural conditions in feed as well as in food. Some of the most common mycotoxins include: aflatoxins, trichothecenes, fumonisins, zearalenone, ochratoxin and ergot alkaloids. Mycotoxins are produced by different strains of fungi and each strain can produce more than one mycotoxin. The major classes of these mycotoxin-producing fungi are listed in Table.

Each plant can be affected by more than one fungus and each fungus can produce more than one mycotoxin. Consequently, there is a high probability that many mycotoxins are present in one feed ingredient, thus increasing the chances of interaction between mycotoxins and the occurrence of synergistic effects, which are of great concern in livestock health and productivity. Synergistic effects occur when the combined effects of two mycotoxins (even at low levels) are greater than the individual effects of each toxin alone. Simple additive effects can also occur with the combined effects of two mycotoxins being equal to the sum of the effects of each toxin on its own.

Mycotoxins are invisible, tasteless, chemically stable and resistant to temperature and storage. They are resistant the normal feed manufacturing processes.

Types of Mycotoxins

Aflatoxins

Aflatoxin: This mycotoxin is primarily produced by Aspergillus species. It is one of the most potent carcinogens known to man and has been linked to a wide array of human health problems. The FDA has established a maximum allowable level of total aflatoxin in food commodities of 20 parts per billion (ppb) and the maximum level for aflatoxin in milk products is 0.5 ppb. Aflatoxin is a potent human carcinogen. It is a naturally occurring toxic metabolite produced by certain fungi (Aspergillus flavis), a mold found on food products such as corn and peanuts, peanut butter. It acts as a potent liver carcinogen in rodents (and, presumably,humans).

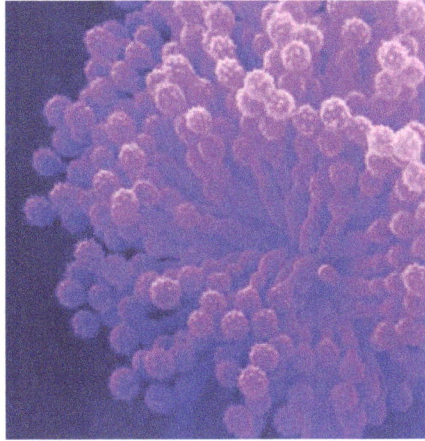

They are probably the best known and most intensively researched mycotoxins in the world. Aflatoxins have been associated with various diseases , such as aflatoxicosis , in livestock , domestic animals and humans throughout the world. The occurrence of aflatoxins is influenced by the weather, (temperature, and humidity - warm & wet is worst!); so the extent of contamination will vary with geographic location , agricultural and agronomic practices, and the susceptibility of the peanuts to fungus before they are harvested, and during storage, and/or processing periods . Aflatoxins have received greater attention than any other mycotoxins because they clearly have a potent carcinogenic effect in laboratory rats and their acute poisonous effects in humans . In the 1960 more than 100,000 young turkeys on poultry farms in England died in the course of a few months from an apparently new disease that was termed "Turkey X disease" . It was soon found that the difficulty was not limited to turkeys . Ducklings and young pheasants were also affected and heavy mortality was experienced.

Ochratoxin

Ochratoxin: This mycotoxin is primarily produced by species of Penicillium and Aspergillus. Ochratoxin is primarily a kidney toxin but if the concentration is sufficiently high there can be damage to the liver as well. Ochratoxin is a carcinogen in rats and mice and is suspect as the causative agent of a human disease, Balkan Endemic Nephropathy, that affects the kidneys and often tumors are associated with the disease. The toxin may be still present in products made from grain and the human population is exposed in this manner.

A significant impact of ochratoxin is that it occurs in such a wide variety of commodities such as raisins, barley, soy products and coffee in amounts that may be relatively low. However, the levels may accumulate in the body of either humans or animals consuming contaminated food because ochratoxin is often not rapidly removed from the body and significant amounts may accumulate in the blood and other selected tissues. The awareness of the occurrence of ochratoxin in this wide variety of commodities has been possible through increased sensitivity of the methods for the analysis of ochratoxin.

Tricothecene

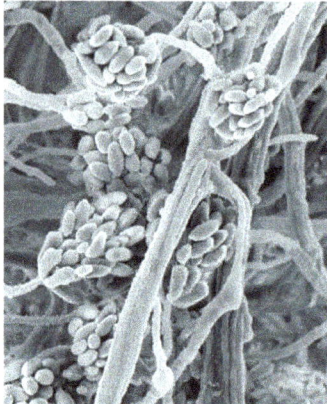

Tricothecene: The toxin is produced by Stachybotrys spp., Fusarium spp and has even been indicated as a potential agent for use as a biological weapon. One of the more deadly mycotoxins, if it is ingested in large amounts it can severely damage the entire digestive tract and cause rapid death due to internal hemorrhaging. It has also been implicated in human disease such as infant pulmonary hemosiderosis. Mycotoxins act by inhibition of protein synthesis. Symptoms start within minutes to hours after exposure, and involve eyes, skin, respiratory and gastrointestinal tracts.

Zearalenone

Zearalenone: A resorcylic acid lactone, is a nonsteroidal estrogenic mycotoxin produced by numerous species of Fusarium. As a result zearalenone is found in a number of cereal crops and their derived food products. A closely related substance "zeranol" (zearalanol) is at present being used in the United States and Canada as an anabolic agent in beef cattle. Zearalenone has been implicated in numerous incidences of mycotoxicosis in farm animals, especially pigs. In this report the health risks to Canadians due to the presence of zearalenone in food products have been evaluated. The first part of the report deals with the physicochemical aspects, mycology, laboratory production, and natural occurrence in plant products and animal products of zearalenone. The stability of zearalenone in foods and feeds, the effects of food processing, and the removal from foods and feeds by physicochemical means are also discussed.

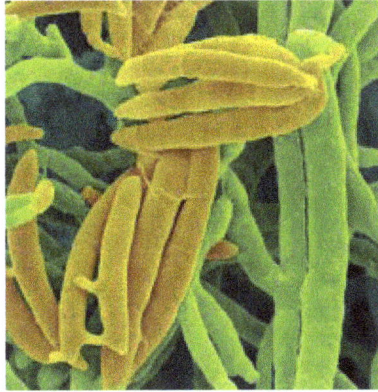

Fumonisins

Fumonisins: are a group of recently characterised mycotoxins produced by F. moniliforme, a mould which occurs worldwide and is frequently found in maize. Fumonisin B1 has been reported in maize (and maize products) from a variety of agroclimatic regions including the USA, Canada, Uruguay, Brazil, South Africa, Austria, Italy and France. The toxins especially occur when maize is grown under warm, dry conditions. Exposure to fumonisin B1 in maize causes leukoencephalomalacia (LEM) in horses and pulmonary oedema in pigs. LEM has been reported in many countries including the USA, Argentina, Brazil, Egypt, South Africa and China. FB1 is also toxic to the central nervous system, liver, pancreas, kidney and lung in a number of animal species.

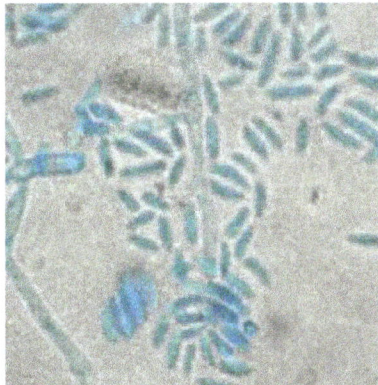

Ethnomycology

Ethnomycology is the study of the historical uses and sociological impact of fungi and can be considered a subfield of ethnobotany or ethnobiology. Although in theory the term includes fungi used for such purposes as tinder, medicine (medicinal mushrooms) and food (including yeast), it is often used in the context of the study of psychoactive mushrooms such as psilocybin mushrooms, the *Amanita muscaria* mushroom, and the ergot fungus.

Amanita muscaria has a long and varied history of psychoactive use.

Lichenology

The branch of mycology that studies lichens is known as lichenology. Lichens are organisms which interact symbiotically with fungi. The study of lichen borrows from many studies of science, including mycology, phycology, microbiology, and botany. Lichenologists is the name of scientists studying lichenology. It has origins going back to, and has influenced many of the afore mentioned studies as much as it borrows from them.

Study of lichens draws knowledge from several disciplines: mycology, phycology, microbiology and botany. Scholars of lichenology are known as lichenologists.

The taxonomy of lichens was first intensively investigated by the Swedish botanist Erik Acharius (1757–1819), who is therefore sometimes named the "father of lichenology". Acharius was a student of Carl Linnaeus. Some of his more important works on the subject, which marked the beginning of lichenology as a discipline, are:

- *Lichenographiae Suecia prodromus* (1798)
- *Methodus lichenum* (1803)

- *Lichenographia universalis* (1810)
- *Synopsis methodica lichenum* (1814)

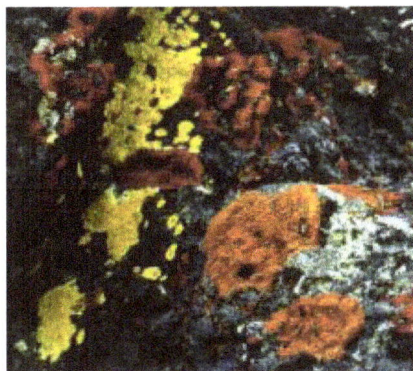
Lichen on rocks

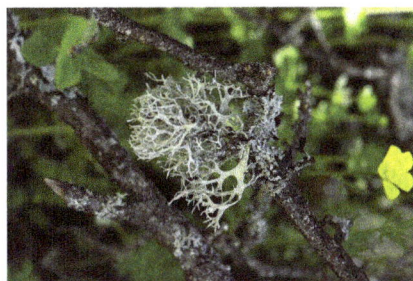
Evernia prunastri, oakmoss

Later lichenologists include the American scientists Vernon Ahmadjian and Edward Tuckerman and the Russian evolutionary biologist Konstantin Merezhkovsky, as well as amateurs such as Louisa Collings

Lichen

A lichen is not a single organism; it is a stable symbiotic association between a fungus and algae and/or cyanobacteria.

Like all fungi, lichen fungi require carbon as a food source; this is provided by their symbiotic algae and/or cyanobacteria, that are photosynthetic.

The lichen symbiosis is thought to be a mutualism, since both the fungi and the photosynthetic partners, called photobionts, benefit.

How do Lichens Grow?

The algal and/or cyanobacterial partner(s) possess the green pigment chlorophyll, enabling them to use sunlight's energy to make their own food from water and carbon dioxide through photosynthesis. They also provide vitamins to the fungus. Cyanobacteria can make amino acids directly from the nitrogen gas in the atmosphere, something neither fungi nor algae can do. The fungus, in turn, protects its partners from drying out and shades them from strong sunlight by enclosing the photosynthesizing partners within the body of the lichen.

This life habit has allowed lichens to successfully colonize many different habitats. Lichens have a truly remarkable resistance to drought. A dry lichen can quickly absorb from 3 to 35 times its weight in water! Lichens can also absorb moisture from dew or fog, even from the air itself if the humidity is very high and the temperature is low. They also dry out slowly, making it possible for the photosynthesizing partner(s) to make food for as long as possible. This ability to quickly absorb and retain water from many sources makes it possible for lichens to live in harsh environments like deserts and polar regions, and on exposed surfaces like bare rocks, roofs and tree branches.

The thallus, or lichen body, comes in four shapes:

- Foliose: flat leaf-like lichens.

- Crustose: crust-like lichens that may be buried in tree bark, or even between the crystals of rocks

- Fruticose: miniature shrub-like lichens.—one lichen of this type is the famous "reindeer moss" of Lapland.

- Squamulose: scaly lichens made of numerous small rounded lobes, intermediate between foliose and crustose lichens.

Most lichens grow slowly, probably because they live in environments where water is available for only short periods. They tend to live for many years, and lichens hundred of years old can be used to date the rock surfaces on which they grow. Lichens spread mostly by small pieces of their body being blown around. All the partners in the original lichen body are present in the fragment, so growth can begin immediately. Some lichens create soredia, balls of tissue made just for dispersal. Although the fungus is the major partner, dispersal by spores is rare.

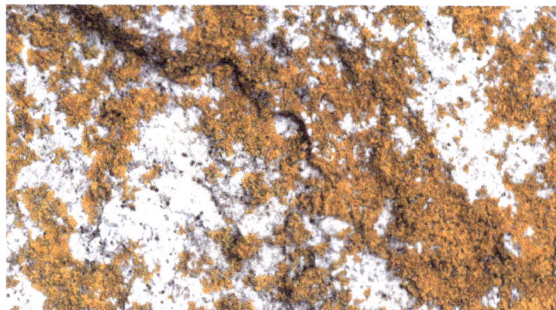

Reproduction

Lichens reproduce either by tiny parts of the lichen breaking off and growing somewhere else, or by the fungal partner producing spores. Lichens may have powdery masses on their surface. These are the tiny bits of the lichen body which will be shed to form new lichens. The individual bits are called soredia and they contain both the fungus and the algal partner together.

In most cases, fungal spores are either produced in apothecia or perithecia on the surface of the lichen. The spores come only from the fungal partner and do not contain any algal cells. They may germinate after being shed from the fruiting body, but they will only be able to form a new lichen if they happen to make contact with a suitable algal partner. Without the alga, the germinating spore will die, as the fungus cannot survive on its own.

The fruiting bodies of most lichens are unusual in that they may continue to produce spores at intervals for several years. The fruiting structures of individual fungi in contrast usually last for a relatively short period.

Taxonomy and Classification

Lichens are classified by the fungal component. Lichen species are given the same scientific name (binomial name) as the fungus species in the lichen. Lichens are being integrated into the classification schemes for fungi. The alga bears its own scientific name, which bears no relationship to that of the lichen or fungus. There are about 13,500–17,000 identified lichen species. Nearly 20% of known fungal species are associated with lichens.

"Lichenized fungus" may refer to the entire lichen, or to just the fungus. This may cause confusion without context. A particular fungus species may form lichens with different algae species, giving rise to what appear to be different lichen species, but which are still classified (as of 2014) as the same lichen species.

Formerly, some lichen taxonomists placed lichens in their own division, the Mycophycophyta, but this practice is no longer accepted because the components belong to separate lineages. Neither the ascolichens nor the basidiolichens form monophyletic lineages in their respective fungal phyla, but they do form several major solely or primarily lichen-forming groups within each phylum. Even more unusual than basidiolichens is the fungus *Geosiphon pyriforme*, a member of the Glomeromycota that is unique in that it encloses a cyanobacterial symbiont inside its cells. *Geosiphon* is not usually considered to be a lichen, and its peculiar symbiosis was not recognized for many years. The genus is more closely allied to endomycorrhizal genera.

Lichens independently emerged from fungi associating with algae and cyanobacteria multiple times throughout history.

Fungi

The fungal component of a lichen is called the mycobiont. The mycobiont may be an Ascomycete or Basidiomycete. The associated lichens are called either ascolichens or basidiolichens, respectively. Living as a symbiont in a lichen appears to be a successful way for a fungus to derive essential nutrients since about 20% of all fungal species have acquired this mode of life.

Thalli produced by a given fungal symbiont with its differing partners may be similar, and the secondary metabolites identical, indicating that the fungus has the dominant role in determining the morphology of the lichen. But the same mycobiont with different photobionts may also produce very different growth forms. Lichens are known in which there is one fungus associated with two or even three algal species.

Although each lichen thallus generally appears homogeneous, some evidence seems to suggest that the fungal component may consist of more than one genetic individual of that species.

Two or more fungal species can interact to form the same lichen.

Photobionts

The photosynthetic partner in a lichen is called a photobiont. The photobionts in lichens come from a variety of simple prokaryotic and eukaryotic organisms. In the majority of lichens the photobiont is a green alga (Chlorophyta) or a cyanobacterium. In some lichens both types are present. Algal photobionts are called phycobionts, while cyanobacterial photobionts are called cyanobionts. According to one source, about 90% of all known lichens have phycobionts, and about 10% have cyanobionts, while another source states that two thirds of lichens have green algae as phycobiont, and about one third have a cyanobiont. Approximately 100 species of photosynthetic partners from 40 genera and five distinct classes (prokaryotic: Cyanophyceae; eukaryotic: Trebouxiophyceae, Phaeophyceae, Chlorophyceae) have been found to associate with the lichen-forming fungi.

Common algal photobionts are from the genus *Trebouxia*, *Trentepohlia*, *Pseudotrebouxia*, or *Myrmecia*. *Trebouxia* is the most common genus of green algae in lichens, occurring in about 40% of all lichens. "Trebouxioid" means either a photobiont that is in the genus *Trebouxia*, or resembles a member of that genus, and is therefore presumably a member of the class Trebouxiophyceae. The second most commonly represented green alga genus is *Trentepohlia*. Overall, about 100 species of eukaryotes are known to occur as photobionts in lichens. All the algae are probably able to exist independently in nature as well as in the lichen.

A "cyanolichen" is a lichen with a cyanobacterium as its main photosynthetic component (photobiont). The most commonly occurring cyanobacterium genus is *Nostoc*. Other common cyanobacterium photobionts are from *Scytonema*. Many cyanolichens are small and black, and have limestone as the substrate. Another cyanolichen group, the jelly lichens of the genera *Collema* or *Leptogium* are gelatinous and live on moist soils. Another group of large and foliose species including *Peltigera*, *Lobaria*, and *Degelia* are grey-blue, especially when dampened or wet. Many of these characterize the Lobarion communities of higher rainfall areas in western Britain, e.g., in the Celtic rain forest. Strains of cyanobacteria found in various cyanolichens are often closely related to one another. They differ from the most closely related free-living strains.

The lichen association is a close symbiosis. It extends the ecological range of both partners but is not always obligatory for their growth and reproduction in natural environments, since many of the algal symbionts can live independently. A prominent example is the alga *Trentepohlia*, which forms orange-coloured populations on tree trunks and suitable rock faces. Lichen propagules (diaspores) typically contain cells from both partners, although the fungal components of so-called "fringe species" rely instead on algal cells dispersed by the "core species".

The same cyanobiont species can occur in association with different fungal species as lichen partners. The same phycobiont species can occur in association with different fungal species as lichen partners. More than one phycobiont may be present in a single thallus.

Although each lichen thallus generally appears homogeneous, some evidence seems to suggest that the photobiont component may consist of more than one genetic individual of that species. A single lichen may contain several algal genotypes. These multiple genotypes may better enable response to adaptation to environmental changes, and enable the lichen to inhabit a wider range of environments.

Controversy over Classification Method and Species Names

There are about 20,000 known lichen species. But what is meant by "species" is different from what is meant by biological species in plants, animals, or fungi, where being the same species implies that there is a common ancestral lineage. Because lichens are combinations of members of two or even three different biological kingdoms, these components *must* have a *different* ancestral lineage from each other. By convention, lichens are still called "species" anyway, and are classified according to the species of their fungus, not the species of the algae or cyanobacteria. Lichens are given the same scientific name (binomial name) as the fungus in them, which may cause some confusion. The alga bears its own scientific name, which has no relationship to the name of the lichen or fungus.

Depending on context, "lichenized fungus" may refer to the entire lichen, or to the fungus when it is in the lichen, which can be grown in culture in isolation from the algae or cyanobacteria. Some algae and cyanobacteria are found naturally living outside of the lichen. The fungal, algal, or cyanobacterial component of a lichen can be grown by itself in culture. When growing by themselves, the fungus, algae, or cyanobacteria have very different properties than those of the lichen. Lichen properties such as growth form, physiology, and biochemistry, are very different from the combination of the properties of the fungus and the algae or cyanobacteria.

The same fungus growing in combination with different algae or cyanobacteria, can produce lichens that are very different in most properties, meeting non-DNA criteria for being different "species". Historically, these different combinations were classified as different species. When the fungus is identified as being the same using modern DNA methods, these apparently different species get reclassified as the *same* species under the current (2014) convention for classification by fungal component. This has led to debate about this classification convention. These apparently different "species" have their own independent evolutionary history.

There is also debate as to the appropriateness of giving the same binomial name to the fungus, and to the lichen that combines that fungus with an alga or cyanobacterium (synecdoche). This is especially the case when combining the same fungus with different algae or cyanobacteria produces dramatically different lichen organisms, which would be considered different species by any measure other than the DNA of the fungal component. If the whole lichen produced by the same fungus growing in association with different algae or cyanobacteria, were to be classified as different "species", the number of "lichen species" would be greater.

Diversity

The largest number of lichenized fungi occur in the Ascomycota, with about 40% of species forming such an association. Some of these lichenized fungi occur in orders with nonlichenized fungi that live as saprotrophs or plant parasites (for example, the Leotiales, Dothideales, and Pezizales). Other lichen fungi occur in only five orders in which all members are engaged in this habit (Orders Graphidales, Gyalectales, Peltigerales, Pertusariales, and Teloschistales). Overall, about 98% of lichens have an ascomycetous mycobiont. Next to the Ascomycota, the largest number of lichenized fungi occur in the unassigned fungi imperfecti, a catch-all category for fungi whose sexual form of reproduction has never been observed. Comparatively few Basidiomycetes are lichenized, but these include agarics, such as species of *Lichenomphalia*, clavarioid fungi, such as species of *Multiclavula*, and corticioid fungi, such as species of *Dictyonema*.

Identification Methods

Lichen identification uses growth form and reactions to chemical tests.

"Pd" refers to the outcome of the Pd test or is used as an abbreviation for the chemical used in the test, para-phenylenediamine. If putting a drop on a lichen turns an area bright yellow to orange, this helps identify it as belonging to either the genus *Cladonia* or *Lecanora*.

Lichens as Bioindicators

Bioindicators are living organisms that respond in an especially clear way to a change in the environment. The hardy lichens are useful bioindicators for air pollution, espeially sulfur dioxide pollution, since they derive their water and essential nutrients mainly from the atmosphere rather than from the soil. It also helps that they are able to react to air pollutants all year round. Compared with most physical/chemical monitors, they are inexpensive to use in evaluating air pollution.

Lichens can also be used to measure toxic elemental pollutants and radioactive metals because they bind these substances in their fungal threads where they concentrate them over time. Environmental scientists can then evaluate this accumulation to determine the history of the local air.

Lichenometry

Lichenometry is a method of numerical dating that uses the size of lichen colonies on a rock surface to determine the surface's age. Lichenometry is used for rock surfaces less than about 10,000 years old.

The basic premise of lichenometry is that the diameter of the largest lichen thallus growing on a moraine, or other surface, is proportional to the length of time that the surface has been exposed to colonisation and growth. Data on lichen growth rates can enable estimates of both the age of the thallus and the period of exposure of a rock surface to be made.

As a field technique it has the advantage that measurements are relatively simple and easy to obtain. Several factors however limit the application of the technique. The need for local date calibration is paramount. Lichens provide a minimum value of how long the rock surface has remained immobile and undisturbed.

References

- Hawksworth DL, Lücking R (July 2017). "Fungal Diversity Revisited: 2.2 to 3.8 Million Species". Microbiology Spectrum. 5 (4): 79. doi:10.1128/microbiolspec.FUNK-0052-2016. ISBN 9781555819576. PMID 28752818

- Brakhage AA (December 2005). "Systemic fungal infections caused by Aspergillus species: epidemiology, infection process and virulence determinants". Current Drug Targets. 6 (8): 875–86. doi:10.2174/138945005774912717. PMID 16375671

- Speer BR, Waggoner B. "Fossil Record of Lichens". University of California Museum of Paleontology. Retrieved 2010-02-16

- Taylor WA, Free CB, Helgemo R, Ochoada J (2004). "SEM analysis of spongiophyton interpreted as a fossil lichen". International Journal of Plant Sciences. 165 (5): 875–881. doi:10.1086/422129

- Purvis W (2000). Lichens. Washington, D.C.: Smithsonian Institution Press in association with the Natural History Museum, London. pp. 49–75. ISBN 1-56098-879-7

- Simon-Nobbe B, Denk U, Pöll V, Rid R, Breitenbach M (2008). "The spectrum of fungal allergy". International Archives of Allergy and Immunology. 145 (1): 58–86. doi:10.1159/000107578. PMID 17709917

- Speer, Brian R; Ben Waggoner (May 1997). "Lichens: Life History & Ecology". University of California Museum of Paleontology. Retrieved 28 April 2015

- Grube, M; Cardinale, M; De Castro Jr, J. V.; Müller, H; Berg, G (2009). "Species-specific structural and functional diversity of bacterial communities in lichen symbioses". The ISME Journal. 3(9): 1105–1115. doi:10.1038/ismej.2009.63. PMID 19554038

- Chang S-T, Miles PG (2004). Mushrooms: Cultivation, Nutritional Value, Medicinal Effect and Environmental Impact. Boca Raton, Florida: CRC Press. ISBN 0-8493-1043-1

- Williams, Thomas A. (1856). "The Status of the Algo-Lichen Hypothesis". The American Naturalist. 23 (265). doi:10.1086/274846

- "Lichens". National Park Service, US Department of the Interior, Government of the United States. 22 May 2016. Retrieved 4 April2018

- Edwards D, Axe L (2012). "Evidence for a fungal affinity for Nematasketum, a close ally of Prototaxites". Botanical Journal of the Linnean Society. 168: 1–18. doi:10.1111/j.1095-8339.2011.01195.x

2

Biological Classification of Fungi

Fungi are eukaryotic heterotrophic organisms that acquire food by absorbing dissolved molecules. Some examples of fungi are mushrooms, molds and yeasts. The major groups of fungi are Chytridiomycota, Blastocladiomycota, Zygomycota, Ascomycota, Glomeromycota and Basidiomycota, which have been discussed in detail in this chapter.

Dikarya

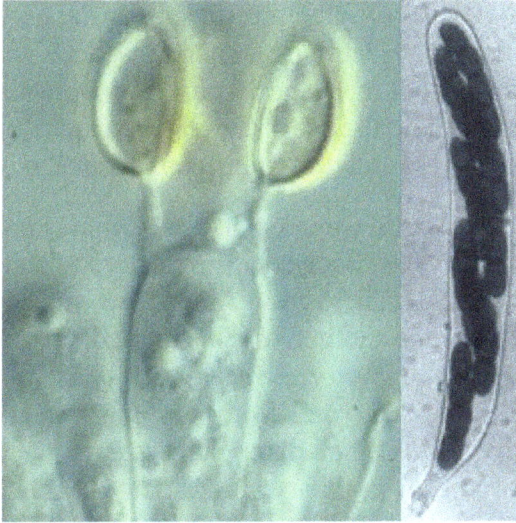

The Dikarya embrace two great phyla, the Ascomycota and the Basidiomycota, accounting for the vast majority of fungi. The obvious difference between these two groups is the way that they produce their spores following meiosis, depicted in the picture at left. The Basidiomycota produce their meiospores, called basidiospores, externally on basidia, while the Ascomycota produce theirs, called ascospores, in asci. The picture does not do justice to the variety of these structures. Basidia can be round or elongated, 1-spores to 8-spored, septate or nonseptate. Asci can be round to elongated and contain a single spore or more than a thousand. Ascospores can range from single-celled to many-celled. In general we can say that basidia are most commonly 4-spored and asci 8-spored, and that basidiospores are almost always single-celled. One invariable fact is that basidiospores are always borne externally on the basidium and ascospores are always borne internally in the ascus.

There are other less obvious differences between the Asco- and Basidiomycota. Of course as members of the Dikarya both groups spend part of their lives in a dikaryotic state, one in which their cells contain two genetically different nuclei that divide together so that the daughter nucleus of each one moves into the new cell.

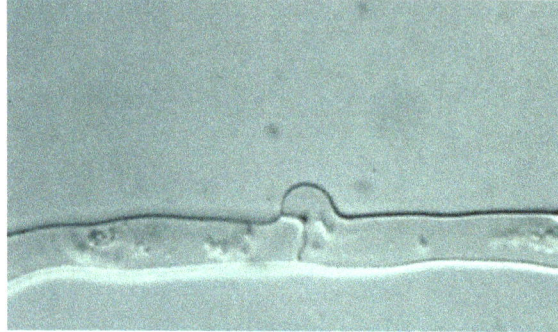

Clamp connections probably occur throughout the Dikarya but are most readily seen in the Basidiomycota. If you can see clamp connections you are most certainly looking at one of the Basidiomycota. The fact that clamp connections are rarely seen in the Ascomycota probably indicates that most species, in common with many Basidiomycota, carry out nuclear division without the use of these structures. However, ascomycetes often produce croziers, structures similar to clamp connections in both form and function.

There is another compelling reason why we don't see clamp connections in the Ascomycota. In the life cycle of most Basidiomycota the dikaryotic state becomes established quite soon after the first hyphae begin to grow. Among growing hyphae many are sexually compatible and pass nuclei to one another, establishing a dikaryon in the process. The resulting dikaryotic hyphae then carry on growing, leading to a large and entirely dikaryotic colony. In the Ascomycota the dikaryon becomes established in specialized cells called ascogonia which then give rise to small dikaryotic systems that are soon converted via meiosis to asci and ascospores. The dikaryotic systems, and their potential clamp connections, are just too limited and too short-lived to be observed easily. So here we have a second fundamental difference between the Ascomycota and the Basidiomycota:

- The Ascomycota have small and usually short-lived dikaryotic states that are dependent upon the monokaryotic state

- The Basidiomycota have large and usually long-live dikaryotic states that are not dependent on the monokaryotic state

Ascomycota

The majority of known fungi belong to the Phylum Ascomycota, which is characterized by the formation of an ascus (plural, asci), a sac-like structure that contains haploid ascospores. Many ascomycetes are of commercial importance. Some play a beneficial role, such as the yeasts used in baking, brewing, and wine fermentation, plus truffles and morels, which are held as gourmet delicacies. *Aspergillus oryzae* is used in the fermentation of rice to produce sake. Other ascomycetes parasitize plants and animals, including humans. For example, fungal pneumonia poses a significant threat to AIDS patients who have a compromised immune system. Ascomycetes not only infest and destroy crops directly; they also produce poisonous secondary metabolites that make crops unfit for consumption. Filamentous ascomycetes produce hyphae divided by perforated septa, allowing streaming of cytoplasm from one cell to the other. Conidia and asci, which are used respectively for asexual and sexual reproductions, are usually separated from the vegetative hyphae by blocked (non-perforated) septa.

Metabolism

In common with other fungal phyla, the Ascomycota are heterotrophic organisms that require organic compounds as energy sources. These are obtained by feeding on a variety of organic substrates including dead matter, foodstuffs, or as symbionts in or on other living organisms. To obtain these nutrients from their surroundings, ascomycetous fungi secrete powerful digestive enzymes that break down organic substances into smaller molecules, which are then taken up into the cell. Many species live on dead plant material such as leaves, twigs, or logs. Several species colonize plants, animals, or other fungi as parasites or mutualistic symbionts and derive all their metabolic energy in form of nutrients from the tissues of their hosts.

Owing to their long evolutionary history, the Ascomycota have evolved the capacity to break down almost every organic substance. Unlike most organisms, they are able to use their own enzymes to digest plant biopolymers such as cellulose or lignin. Collagen, an abundant structural protein in animals, and keratin—a protein that forms hair and nails—, can also serve as food sources. Unusual examples include *Aureobasidium pullulans*, which feeds on wall paint, and the kerosene fungus *Amorphotheca resinae*, which feeds on aircraft fuel (causing occasional problems for the airline industry), and may sometimes block fuel pipes. Other species can resist high osmotic stress and grow, for example, on salted fish, and a few ascomycetes are aquatic.

The Ascomycota is characterized by a high degree of specialization; for instance, certain species of Laboulbeniales attack only one particular leg of one particular insect species. Many Ascomycota engage in symbiotic relationships such as in lichens—symbiotic associations with green algae or cyanobacteria—in which the fungal symbiont directly obtains products of photosynthesis. In common with many basidiomycetes and Glomeromycota, some ascomycetes form symbioses with plants by colonizing the roots to form mycorrhizal associations. The Ascomycota also represents several carnivorous fungi, which have developed hyphal traps to capture small protists such as amoebae, as well as roundworms (*Nematoda*), rotifers, tardigrades, and small arthropods such as springtails (*Collembola*).

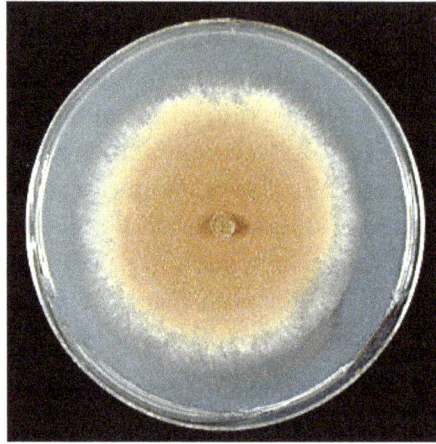

Hypomyces completus on culture medium

Distribution and Living Environment

The Ascomycota are represented in all land ecosystems worldwide, occurring on all continents including Antarctica. Spores and hyphal fragments are dispersed through the atmosphere and freshwater environments, as well as ocean beaches and tidal zones. The distribution of species is variable; while some are found on all continents, others, as for example the white truffle *Tuber magnatum*, only occur in isolated locations in Italy and Eastern Europe. The distribution of plant-parasitic species is often restricted by host distributions; for example, *Cyttaria* is only found on *Nothofagus* (Southern Beech) in the Southern Hemisphere.

Reproduction

Asexual Reproduction

Asexual reproduction is the dominant form of propagation in the Ascomycota, and is responsible for the rapid spread of these fungi into new areas. It occurs through vegetative reproductive spores, the conidia. The conidiospores commonly contain one nucleus and are products of mitotic cell divisions and thus are sometimes called mitospores, which are genetically identical to the mycelium from which they originate. They are typically formed at the ends of specialized hyphae, the *conidiophores*. Depending on the species they may be dispersed by wind or water, or by animals.

Asexual Spores

Different types of asexual spores can be identified by colour, shape, and how they are released as individual spores. Spore types can be used as taxonomic characters in the classification within the Ascomycota. The most frequent types are the single-celled spores, which are designated *amerospores*. If the spore is divided into two by a cross-wall (septum), it is called a *didymospore*.

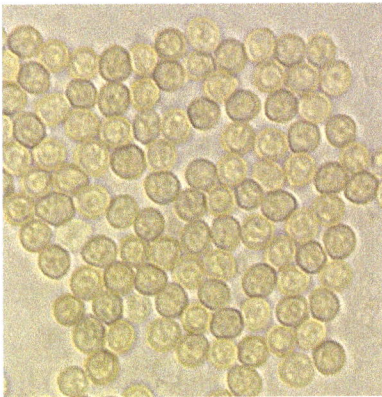

Conidiospores of *Trichoderma aggres-sivum*, Diameter approx. 3µm

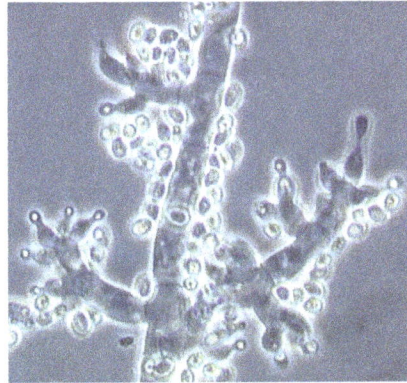

Conidiophores of molds of the genus *Aspergillus*, conidiogenesis is blastic-phialidic

When there are two or more cross-walls, the classification depends on spore shape. If the septae are *transversal*, like the rungs of a ladder, it is a *phragmospore*, and if they possess a net-like structure it is a *dictyospore*. In *staurospores* ray-like arms radiate from a central body; in others (*helicospores*) the entire spore is wound up in a spiral like a spring. Very long worm-like spores with a length-to-diameter ratio of more than 15:1, are called *scolecospores*.

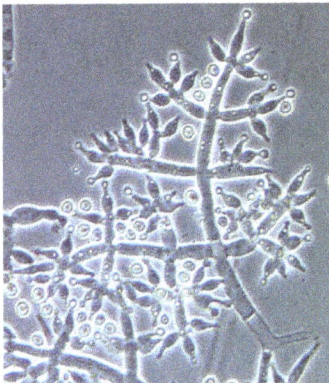

Conidiophores of Trichoderma harzianum, conidiogenesis is blas-tic-phialidic

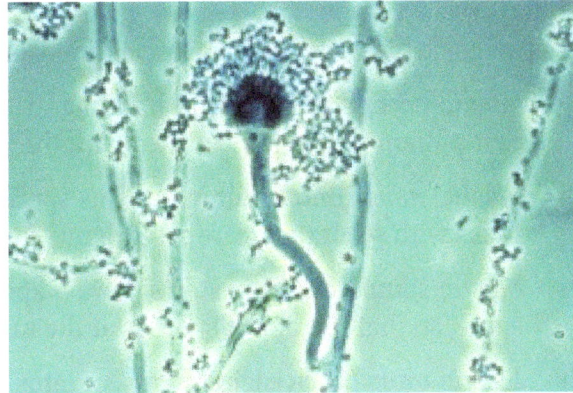

Conidiophores of Trichoderma fertile with vase-shaped phialides and newly formed conidia on their ends (bright points)

Conidiogenesis and Dehiscence

Important characteristics of the anamorphs of the Ascomycota are *conidiogenesis*, which includes spore formation and dehiscence (separation from the parent structure). Conidiogenesis corresponds to Embryology in animals and plants and can be divided into two fundamental forms of development: *blastic* conidiogenesis, where the spore is already evident before it separates from the conidiogenic hypha, and *thallic* conidiogenesis, during which a cross-wall forms and the newly created cell develops into a spore. The spores may or may not be generated in a large-scale specialized structure that helps to spread them.

These two basic types can be further classified as follows:

- *blastic-acropetal* (repeated budding at the tip of the conidiogenic hypha, so that a chain of spores is formed with the youngest spores at the tip),

- *blastic-synchronous* (simultaneous spore formation from a central cell, sometimes with secondary acropetal chains forming from the initial spores),

- *blastic-sympodial* (repeated sideways spore formation from behind the leading spore, so that the oldest spore is at the main tip),

- *blastic-annellidic* (each spore separates and leaves a ring-shaped scar inside the scar left by the previous spore),

- *blastic-phialidic* (the spores arise and are ejected from the open ends of special conidiogenic cells called phialides, which remain constant in length),

- *basauxic* (where a chain of conidia, in successively younger stages of development, is emitted from the mother cell),

- *blastic-retrogressive* (spores separate by formation of crosswalls near the tip of the conidiogenic hypha, which thus becomes progressively shorter),

- *thallic-arthric* (double cell walls split the conidiogenic hypha into cells that develop into short, cylindrical spores called *arthroconidia*; sometimes every second cell dies off, leaving the arthroconidia free),

- *thallic-solitary* (a large bulging cell separates from the conidiogenic hypha, forms internal walls, and develops to a *phragmospore*).

Sometimes the conidia are produced in structures visible to the naked eye, which help to distribute the spores. These structures are called "conidiomata" (singular: conidioma), and may take the form of *pycnidia* (which are flask-shaped and arise in the fungal tissue) or *acervuli* (which are cushion-shaped and arise in host tissue).

Dehiscence happens in two ways. In *schizolytic* dehiscence, a double-dividing wall with a central lamella (layer) forms between the cells; the central layer then breaks down thereby releasing the spores. In *rhexolytic* dehiscence, the cell wall that joins the spores on the outside degenerates and releases the conidia.

Heterokaryosis and Parasexuality

Several Ascomycota species are not known to have a sexual cycle. Such asexual species may be able to undergo genetic recombination between individuals by processes involving *heterokaryosis* and *parasexual* events.

Parasexuality refers to the process of heterokaryosis, caused by merging of two hyphae belonging to different individuals, by a process called *anastomosis*, followed by a series of events resulting in genetically different cell nuclei in the mycelium. The merging of nuclei is not followed by meiotic events, such as gamete formation and results in an increased number of chromosomes per nuclei. *Mitotic crossover* may enable recombination, i.e., an exchange of genetic material between homologous chromosomes. The chromosome number may then be restored to its haploid state by nuclear division, with each daughter nuclei being genetically different from the original parent nuclei. Alternatively, nuclei may lose some chromosomes, resulting in aneuploid cells. *Candida albicans* (class Saccharomycetes) is an example of a fungus that has a parasexual cycle.

Sexual Reproduction

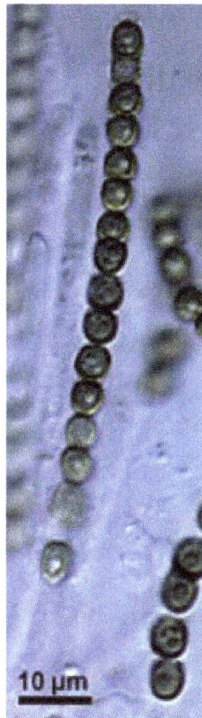

Ascus of *Hypocrea virens* with eight two-celled Ascospores

Sexual reproduction in the Ascomycota leads to the formation of the *ascus*, the structure that defines this fungal group and distinguishes it from other fungal phyla. The ascus is a tube-shaped vessel, a *meiosporangium*, which contains the sexual spores produced by meiosis and which are called *ascospores*.

Apart from a few exceptions, such as *Candida albicans*, most ascomycetes are haploid, i.e., they contain one set of chromosomes per nucleus. During sexual reproduction there is a diploid phase, which commonly is very short, and meiosis restores the haploid state. The sexual cycle of one well-studied representative species of Ascomycota is described in greater detail in Neurospora crassa.

Formation of Sexual Spores

The sexual part of the life cycle commences when two hyphal structures mate. In the case of *homothallic* species, mating is enabled between hyphae of the same fungal clone, whereas in *heterothallic* species, the two hyphae must originate from fungal clones that differ genetically, i.e., those that are of a different mating type. Mating types are typical of the fungi and correspond roughly to the sexes in plants and animals; however one species may have more than two mating types, resulting in sometimes complex vegetative incompatibility systems. The adaptive function of mating type is discussed in Neurospora crassa.

Gametangia are sexual structures formed from hyphae, and are the generative cells. A very fine hypha, called trichogyne emerges from one gametangium, the *ascogonium*, and merges with a gametangium (the *antheridium*) of the other fungal isolate. The nuclei in the antheridium then migrate into the ascogonium, and plasmogamy—the mixing of the cytoplasm—occurs. Unlike in animals and plants, plasmogamy is not immediately followed by the merging of the nuclei (called *karyogamy*). Instead, the nuclei from the two hyphae form pairs, initiating the *dikaryophase* of the sexual cycle, during which time the pairs of nuclei synchronously divide. Fusion of the paired nuclei leads to mixing of the genetic material and recombination and is followed by meiosis. A similar sexual cycle is present in the blue green algae (Rhodophyta). A discarded hypothesis held that a second karyogamy event occurred in the ascogonium prior to ascogeny, resulting in a tetraploid nucleus which divided into four diploid nuclei by meiosis and then into eight haploid nuclei by a supposed process called brachymeiosis, but this hypothesis was disproven in the 1950s.

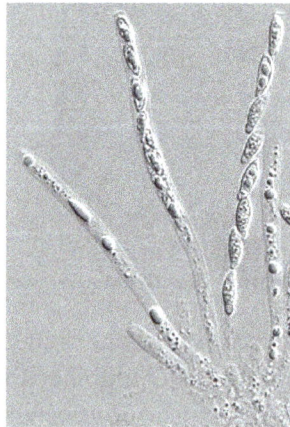

Unitunicate-inoperculate Asci of *Hypomyces chrysospermus*

From the fertilized ascogonium, *dinucleate* hyphae emerge in which each cell contains two nuclei. These hyphae are called *ascogenous* or fertile hyphae. They are supported by the vegetative mycelium containing uni– (or mono–) nucleate hyphae, which are sterile. The mycelium containing both sterile and fertile hyphae may grow into fruiting body, the *ascocarp*, which may contain millions of fertile hyphae.

The sexual structures are formed in the fruiting layer of the ascocarp, the hymenium. At one end of ascogenous hyphae, characteristic U-shaped hooks develop, which curve back opposite to the growth direction of the hyphae. The two nuclei contained in the apical part of each hypha divide in such a way that the threads of their mitotic spindles run parallel, creating two pairs of genetically different nuclei. One daughter nucleus migrates close to the hook, while the other daughter nucleus locates to the basal part of the hypha. The formation of two parallel cross-walls then divides the hypha into three sections: one at the hook with one nucleus, one at the basal of the original hypha that contains one nucleus, and one that separates the U-shaped part, which contains the other two nuclei.

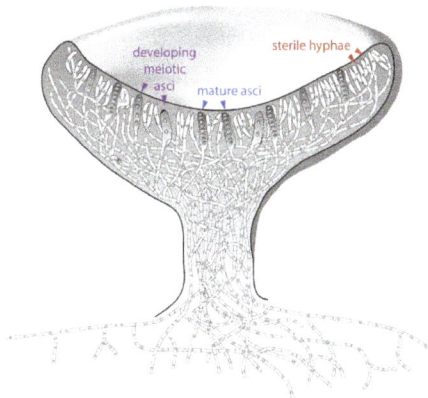

Diagram of an apothecium (the typical cup-like reproductive structure of Ascomycetes) showing sterile tissues as well as developing and mature asci.

Fusion of the nuclei (karyogamy) takes place in the U-shaped cells in the hymenium, and results in the formation of a diploid zygote. The zygote grows into the ascus, an elongated tube-shaped or cylinder-shaped capsule. Meiosis then gives rise to four haploid nuclei, usually followed by a further mitotic division that results in eight nuclei in each ascus. The nuclei along with some cytoplasma become enclosed within membranes and a cell wall to give rise to ascospores that are aligned inside the ascus like peas in a pod.

Upon opening of the ascus, ascospores may be dispersed by the wind, while in some cases the spores are forcibly ejected form the ascus; certain species have evolved spore cannons, which can eject ascospores up to 30 cm. away. When the spores reach a suitable substrate, they germinate, form new hyphae, which restarts the fungal life cycle.

The form of the ascus is important for classification and is divided into four basic types: unitunicate-operculate, unitunicate-inoperculate, bitunicate, or prototunicate.

Cell Structure and Metabolism

The bodies of Ascomycota are eukaryotic cells surrounded by a wall consisting of chitin and beta glucans. They can be single-celled (yeasts) or filamentous (hyphal) organisms. In addition, they can also be dimorphic. The yeasts grow by budding or fission, while hyphae branch out. Most are haploid, but some can be diploid. Spores are stored in cases (asci), which release clouds of spore smoke. Nuclear fusion and meiosis take place within the ascus.

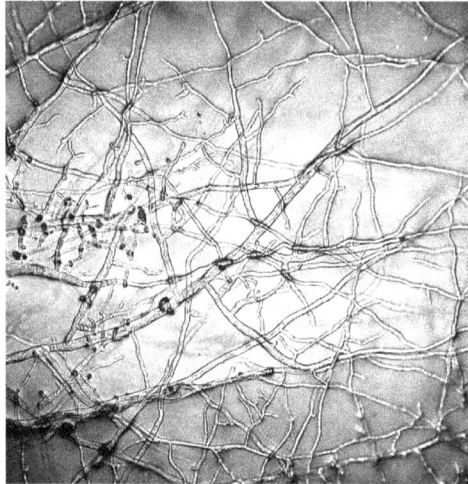

The mycelium—a network of filaments called "hyphae"— is the primary form of a multicellular fungus. This image is of a mycelium growing in an agar-substitute. This image covers a one-millimeter square. Photograph by Bob Blaylock.

Ascomycota are heterotrophic, obtaining nutrients from both dead or living organisms. In addition, these fungi are capable of consuming almost any liquid, as long as there is water present in it.

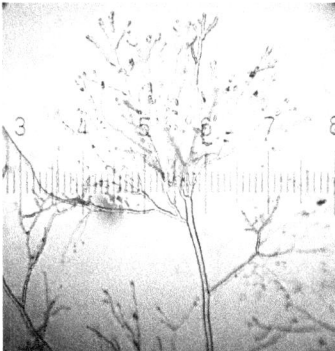

Another mycelium, growing in an agar-substitute. Numbered ticks are 230 µM apart. Photograph by Bob Blaylock.

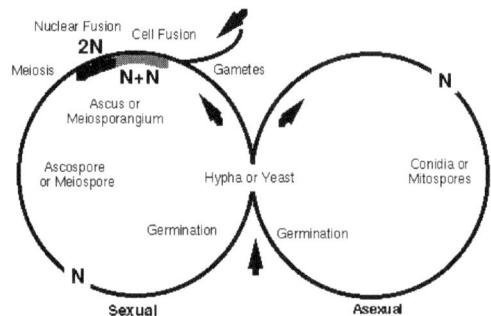

Diagram of life cycle. Ascomycota by John W. Taylor, Joey Spatafora, and Mary Berbee.

Ascomycota have more than one reproductive option. Sexual reproduction takes place within ascospores or meiospores, and asexually with conidia or meitospores. Some must outbreed (heterothallic), some must self breed (homothallic), and some can do both. Meitospores usually simply reproduce the parent organism, but can also act as gametes for fertilization. Reproduction takes place in the ascus, with one round of mitosis following meiosis. This leaves eight nuclei and eight ascospores.

Ecology

Xylaria hypoxylon. Natural Perspective by Ari Kornfeld.

Although certain species live in very specific locations, Ascomycota can be found on all continents. They often form symbiotic relationships with algae, plant roots, and the leaves or stems of plants. However, they do not limit these relationships to plant organisms; they are also known to form them with arthropods. While Ascomycota includes many useful organisms such as fission yeast, baker's yeast, morels, and truffles, it can also account for most animal and plant pathogens. Ascomycota contains the species *Magnaporthe grisea*, which is considered the most destructive pathogen of rice. Dean et. al. (2005) drafted a sequence of this specie's genome in an attempt to better understand its infectious qualities. Ascomycota play a large role in recycling dead plant material.

Many Ascomycota species are useful beyond their conventional functions. Some species, particularly yeasts, have been used in genetics research. For example, a recent experiment by Conrad et. al. (2005) used *Saccharomyces cervisae* to study DNA replication. Other researchers, such as R. Kellermayer (2005) have noted the ways in which *Saccharomyces cervisae* is useful in studying genetic mutations within the context of Hailey-Hailey disease. Cruyssen et. al. (2005) have used this species in their work on Crohn's disease.

The saprophytic Ascomycetes such as yeasts, blue molds and green molds which are cosmopolitan in their distribution are of great value to man. They play an important role in the production of food and medicine.

The yeasts ferment sugar to produce alcohol and carbon dioxide. The entire brewing industry (wine and liquor making) and baking industry (bread making) are built around this simple plant. Food yeast is important as diet supplement.

Yeasts are also a source of many vitamins. Species of Penicillium are employed in the production of cheese and high grade citric acid. Some species of blue and green molds are employed in the production of many other important organic acids.

Some Ascomycetes yield antibiotic drugs such as penicillin and flavicin. Production of penicillin from Penicillium notatum and P. chrysogenum has become a large industry in India and other countries.

Ergot, a valuable drug, is obtained from an ascomycete (Claviceps purpurea) which infects cereals. The drug is used at child birth because it stops bleeding and checks other uterine disturbances.

Included in the Ascomycetes are the popular edible fungi, the truffles and morels, which are considered table delicacies. Neurospora has become an important experimental organism in genetic research.

The negative role the Ascomycetes play in the economy of nature is equally significant. Some spoil foods and other consumer goods. The celluloytic Ascomycetes destroy fabrics containing cellulose.

Some of the parasitic species cause serious diseases of our economic plants. Powdery mildews such as Erysiphe graminis infect various grains. Erysiphe cichoracearum affects fruits and vegetables causing loss by low production.

Peach leaf curl, brown rot of stone fruits, black rot of plums and bitter root of apples are the common diseases of fruit trees caused by parasitic Ascomycetes. The parasitic Ascomycetes also include several severe though uncommon human pathogens.

Several species are the causative agents of diseases of ear and lung. Some of them cause a respiratory disease called aspergillosis. Body ring worm is another fungal disease caused by an ascomycete.

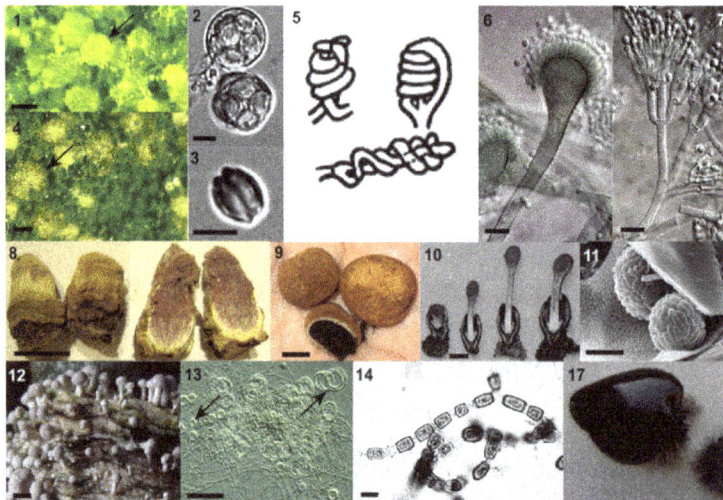

Harmful Interactions

One of their most harmful roles is as the agent of many plant diseases. For instance:

- Dutch Elm Disease, caused by the closely related species *Ophiostoma ulmi* and *Ophiostoma novo-ulmi*, has led to the death of many elms in Europe and North America.

Claviceps purpurea on rye (*Secale cereale*)

- The originally Asian *Cryphonectria parasitica* is responsible for attacking Sweet Chestnuts (*Castanea sativa*), and virtually eliminated the once-widespread American Chestnut (*Castanea dentata*),

- A disease of maize (*Zea mays*), which is especially prevalent in North America, is brought about by *Cochliobolus heterostrophus*.

- *Taphrina deformans* causes leaf curl of peach.

- *Uncinula necator* is responsible for the disease powdery mildew, which attacks grapevines.

- Species of *Monilinia* cause brown rot of stone fruit such as peaches (*Prunus persica*) and sour cherries (*Prunus ceranus*).

- Members of the Ascomycota such as *Stachybotrys chartarum* are responsible for fading of woollen textiles, which is a common problem especially in the tropics.

- Blue-green, red and brown molds attack and spoil foodstuffs - for instance *Penicillium italicum* rots oranges.

- Cereals infected with *Fusarium graminearum* contain mycotoxins like deoxynivalenol (DON), which can lead to skin and mucous membrane lesions when eaten by pigs.

- Ergot (*Claviceps purpurea*) is a direct menace to humans when it attacks wheat or rye and produces highly poisonous and carcinogenic alkaloids, causing ergotism if consumed. Symptoms include hallucinations, stomach cramp, and a burning sensation in the limbs ("Saint Anthony's Fire").

- *Aspergillus flavus*, which grows on peanuts and other hosts, generates aflatoxin, which damages the liver and is highly carcinogenic.

- *Candida albicans*, a yeast that attacks the mucous membranes, can cause an infection of the mouth or vagina called thrush or candidiasis, and is also blamed for "yeast allergies".

- Fungi like *Epidermophyton* cause skin infections but are not very dangerous for people with healthy immune systems. However, if the immune system is damaged they can be life-threatening; for instance, *Pneumocystis jirovecii* is responsible for severe lung infections that occur in AIDS patients.

Ascus

Ascus, plural asci, a saclike structure produced by fungi of the phylum Ascomycota (sac fungi) in which sexually produced spores (ascospores), usually four or eight in number, are formed. Asci may arise from the fungal mycelium(the filaments, or hyphae, constituting the organism) without a distinct fruiting structure, as in the leaf curl fungi; it may arise within a fruiting structure (ascocarp) that may be exposed, as in the molds and powdery mildew fungi; or it may be imbedded in a compact structure (stroma), as in the ergot and black knot fungi. In the case of yeasts, a single cell converts to an ascus.

Ascus classification

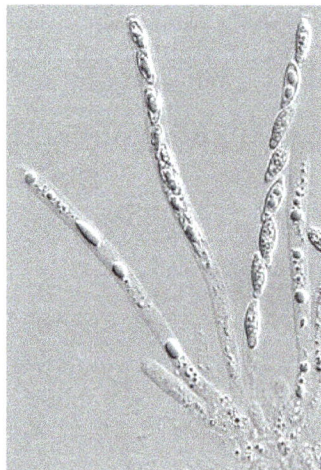

Asci of *Hypomyces chrysospermus* (they are unitunicate-inoperculate). DIC image.

The form of the ascus, the capsule which contains the sexual spores, is important for classification of the Ascomycota. There are four basic types of ascus.

- A unitunicate-operculate ascus has a "lid", the Operculum, which breaks open when the spores are mature and allows the spores to escape. Unitunicate-operculate asci only occur in those ascocarps which have apothecia, for instance the morels. 'Unitunicate' means 'single-walled'.

- Instead of an operculum, a unitunicate-inoperculate ascus has an elastic ring that functions like a pressure valve. Once mature the elastic ring briefly expands and lets the spores shoot out. This type appears both in apothecia and in perithecia; an example is the illustrated *Hypomyces chrysospermus*.

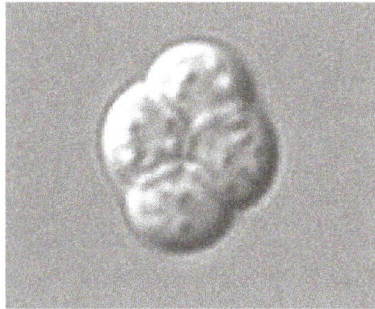

Ascus of *Saccharomyces cerevisiae* containing a tetrad of four spores

- A bitunicate ascus is enclosed in a double wall. This consists of a thin, brittle outer shell and a thick elastic inner wall. When the spores are mature, the shell splits open so that the inner wall can take up water. As a consequence this begins to extend with its spores until it protrudes above the rest of the ascocarp so that the spores can escape into free air without being obstructed by the bulk of the fruiting body. Bitunicate asci occur only in pseudothecia and are found only in the classes *Dothideomycetes* and *Chaetothyriomycetes* (which were formerly united in the old class *Loculoascomycetes*). Examples: *Venturia inaequalis* (apple scab) and *Guignardia aesculi* (Brown Leaf Mold of Horse Chestnut).

- Prototunicate asci are mostly spherical in shape and have no mechanism for forcible dispersal. The mature ascus wall dissolves allowing the spores to escape, or it is broken open by other influences, such as animals. Asci of this type can be found both in perithecia and in cleistothecia, for instance with Dutch elm disease (*Ophiostoma*). This is something of a catch-all term for cases which do not fit into the other three ascus types, and they probably belong to several independent groups which evolved separately from unitunicate asci.

Ascospore

The complex fruiting bodies (ascocarps) of lichen fungi are of several types. The factors that induce fruiting in lichens have not been established with certainty. Spores of lichen fungi (ascospores) are of extremely varying sizes and shapes; e.g., *Pertusaria* has one or two large spores in one ascus (saclike bodies containing the ascospores), and *Acarospora* may have several hundred small spores per ascus. Although in most species the ascospore generally has one nucleus, it may be single-celled or multicellular, brown or colourless; the *Pertusaria* spore, however, is a single cell containing 200 nuclei. Another type of fungal spore may be what are sometimes called spermatia (male fungal sex cells) or pycnidiospores; it is not certain that these structures have the ability to germinate and develop into a fungal colony. Few lichen fungi produce conidia, a type of asexual spore common among ascomycetes.

Basidiomycota

Basidiomycota, large and diverse phylum of fungi (kingdom Fungi) that includes jelly and shelf fungi; mushrooms, puffballs, and stinkhorns; certain yeasts; and the rusts and smuts. Basidiomycota are typically filamentous fungi composed of hyphae. Most species reproduce sexually with a club-shaped spore-bearing organ (basidium) that usually produces four sexual spores(basidiospores). Basidia are borne on fruiting bodies (basidiocarps), which are large and conspicuous in all but the yeasts, rusts, and smuts.

The fungal group basidiomycota is best known for the production of large fruitbodies such as the mushrooms, puffballs, brackets, etc. However, the group also contains some microscopic fungi, including the important rust fungi and smut fungi that parasitise plants, and some yeasts. One of these yeasts, *Sporobolomyces roseus*, is very common on moribund leaf surfaces and is a cause for concern because its basidiospores are respiratory allergens. Another is *Cryptococcus neoformans*, which grows commonly on old "weathered" bird droppings, and which can cause a fatal systemic infection of immunocompromised people. Its air-borne basidiospores initiate infection via the lungs, leading to the disease termed cryptococcosis. This is now the fourth most important life-threatening diseases of AIDS patients in the USA, and common also in Europe.

Many of the basidiomycota with the larger fruitbodies (toadstools etc.) are common and important agents of wood decay or decomposers of leaf litter, animal dung, etc. Others are mycorrhizal on forest trees. Several species are grown commercially for food, including the common cultivated mushroom, *Agaricus bisporus*, and some more exotic species which can be found on supermarket shelves. For example, the supermarket package below contains the brown-coloured Shiitake mushroom (*Lentinus edodes*, traditionally grown in south-east Asia) and a mixture of grey, pink and yellow forms of the oyster fungus (*Pleurotus ostreatus*, reportedly an aphrodisiac).

Distinctive Features and Life Cycle of the Basidiomycota

As a group, the basidiomycota have some highly characteristic features, which separate them from other fungi. They are the most evolutionarily advanced fungi, and even their hyphae have a dinstictly "cellular" composition. This point is best illustrated by the life cycle below.

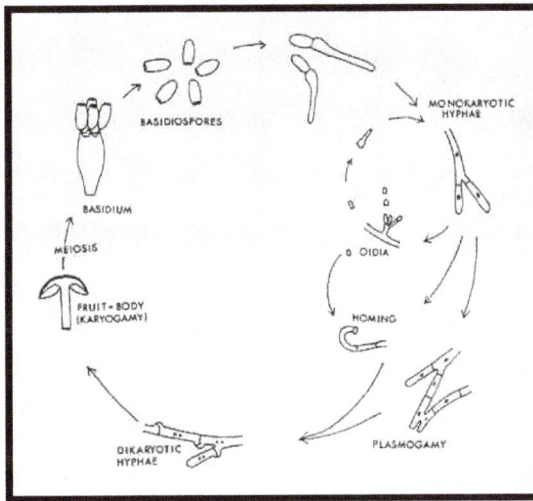

The basidiospores usually have a single haploid nucleus. When these spores germinate they produce hyphae with usually a single nucleus in each compartment. Because these hyphae have only one nuclear type they are termed monokaryons (from the Greek, mono = one; karyos = kernel or nucleus). At some stage in their growth, two monokarons of different compatibility groups (mating types) fuse. This can occur either by simple hyphal fusions or by fusion of a hypha with a small spore termed an oidium. The fusion event is termed plasmogamy. Then the nuclei divide and the daughter nuclei pair so that each hyphal compatment comes to contain two nuclei - one of each mating type. At this stage the fungus is termed a dikaryon (i.e. with two nuclear types).

Many basidiomycota grow for most of their lives as dikaryons, until environmental signals induce them to produce fruibodies, such as a toadstool. All the hyphae that make up the toadstool are dikaryotic. At a late stage of development, some of these hyphae produce special cells termed basidia (singular, basidium). For example, the cells that line the gills of the common mushroom are basidia. Finally, the two haploid nuclei in each basidium fuse - a process termed karyogamy) to form a diploid nucleus. This then undergoes meiosis to produce four haploid nuclei, and these haploid nuclei migrate into the basidiospores, which develop on small stalks (termed sterigmata) from each basidium. The dispersal of these monokaryotic spores completes the life cycle.

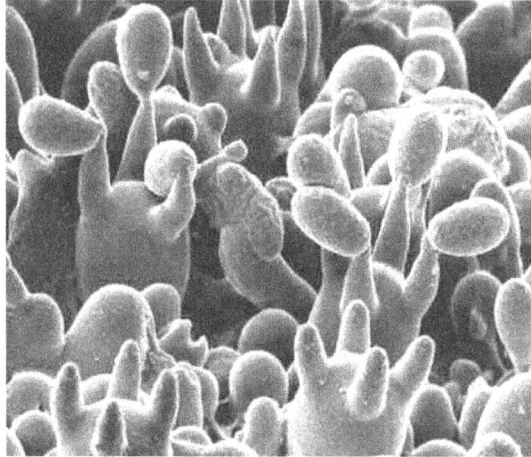

Scanning electron micrograph of basidia on the gills of a toadstool. Note that each basidium produces four stalks (sterigmata) and the basidiospores develop on the ends of these stalks

In many basidiomycota there is a rather elaborate mechanism for ensuring that the dikaryotic condition is maintained during growth of the hyphae. As shown in the diagram below, the two nuclei in each tip cell divide at the same time, but one divides along the axis of the hypha, and the other divides so that a daughter nucleus enters a small, backwards-projecting branch. Then a septum forms to separate the original apical cell into two cells, and the branch fuses with the sub-terminal cell so that the nucleus in the branch migrates into this cell. The small branches at each septum are termed clamp connections. They can be seen at high magnification of a normal compound microscope.

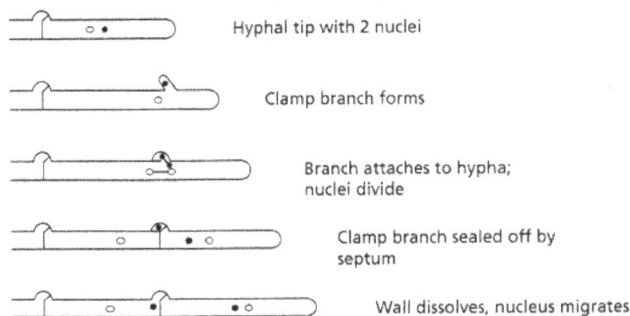

Hyphal tip with 2 nuclei

Clamp branch forms

Branch attaches to hypha; nuclei divide

Clamp branch sealed off by septum

Wall dissolves, nucleus migrates

Diagram to show the role of clamp connections in maintaining a dikaryon

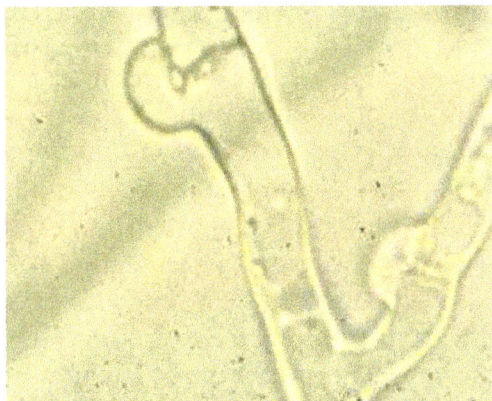

Clamp connections at the septa of a basidiomycota hypha

Typical Life-cycle

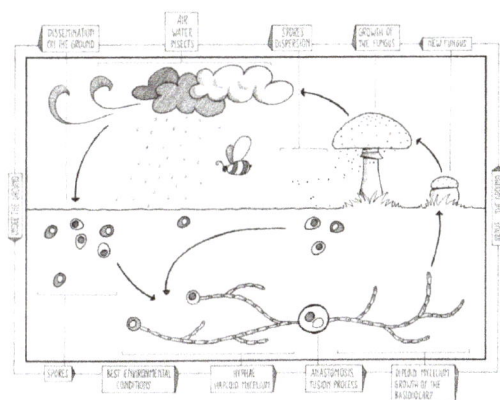

Sexual reproduction cycle of basidiomycetes

Unlike animals and plants which have readily recognizable male and female counter-parts, Basidiomycota (except for the Rust (Pucciniales)) tend to have mutually indistin-guishable, compatible haploids which are usually mycelia being composed of filamen-tous hyphae. Typically haploid Basidiomycota mycelia fuse via plasmogamy and then the compatible nuclei migrate into each other's mycelia and pair up with the resident nuclei. Karyogamy is delayed, so that the compatible nuclei remain in pairs, called a dikaryon. The hyphae are then said to be dikaryotic. Conversely, the haploid mycelia are called monokaryons. Often, the dikaryotic mycelium is more vigorous than the in-dividual monokaryotic mycelia, and proceeds to take over the substrate in which they are growing. The dikaryons can be long-lived, lasting years, decades, or centuries. *The monokaryons are neither male nor female.* They have either a bipolar (unifactorial) or a tetrapolar (bifactorial) mating system. This results in the fact that following meiosis, the resulting haploid basidiospores and resultant monokaryons, have nuclei that are compatible with 50% (if bipolar) or 25% (if tetrapolar) of their sister basidiospores (and their resultant monokaryons) because the mating genes must differ for them to be compatible. However, there are sometimes more than two possible alleles for a given locus, and in such species, depending on the specifics, over 90% of monokaryons could compatible with each other.

The maintenance of the dikaryotic status in dikaryons in many Basidiomycota is facilitated by the formation of clamp connections that physically appear to help coordinate and re-establish pairs of compatible nuclei following synchronous mitotic nuclear divisions. Variations are frequent and multiple. In a typical Basidiomycota lifecycle the long lasting dikaryons periodically (seasonally or occasionally) produce basidia, the specialized usually club-shaped end cells, in which a pair of compatible nuclei fuse (karyogamy) to form a diploid cell. Meiosis follows shortly with the production of 4 haploid nuclei that migrate into 4 external, usually apical basidiospores. Variations occur, however. Typically the basidiospores are ballistic, hence they are sometimes also called ballistospores. In most species, the basidiospores disperse and each can start a new haploid mycelium, continuing the lifecycle. Basidia are microscopic but they are often produced on or in multicelled large fructifications called basidiocarps or basidiomes, or fruitbodies), variously called mushrooms, puffballs, etc. Ballistic basidiospores are formed on sterigmata which are tapered spine-like projections on basidia, and are typically curved, like the horns of a bull. In some Basidiomycota the spores are not ballistic, and the sterigmata may be straight, reduced to stubbs, or absent. The basidiospores of these non-ballistosporic basidia may either bud off, or be released via dissolution or disintegration of the basidia.

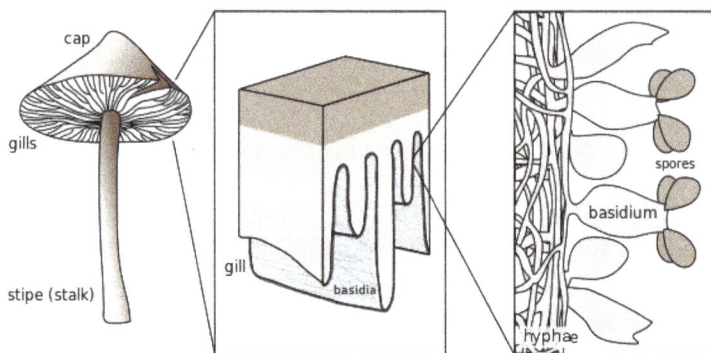

Scheme of a typical basidiocarp, the dipoid reproductive structure of a basidiomycete, showing fruiting body, hymenium and basidia.

In summary, meiosis takes place in a diploid basidium. Each one of the four haploid nuclei migrates into its own basidiospore. The basidiospores are ballistically discharged and start new haploid mycelia called monokaryons. There are no males or females, rather there are compatible thalli with multiple compatibility factors. Plasmogamy between compatible individuals leads to delayed karyogamy leading to establishment of a dikaryon. The dikaryon is long lasting but ultimately gives rise to either fruitbodies with basidia or directly to basidia without fruitbodies. The paired dikaryon in the basidium fuse (i.e. karyogamy takes place). The diploid basidium begins the cycle again.

Meiosis

Coprinopsis cinerea is a multicellular basidiomycete mushroom. It is particularly suited to the study of meiosis because meiosis progresses synchronously in about 10 million cells within the mushroom cap, and the meiotic prophase stage is prolonged. Burns et al. studied the expression of genes involved in the 15-hour meiotic process, and found that the pattern of gene expression of *C. cinerea* was similar to two other fungal species, the yeasts *Saccharomyces cerevisiae* and *Schizosaccharomyces pombe*. These similarities in the patterns of expression led to the conclusion that the core expression program of meiosis has been conserved in these fungi for over half a billion years of evolution since these species diverged.

Cryptococcus neoformans and *Ustilago maydis* are examples of pathogenic basidiomycota. Such pathogens must be able to overcome the oxidative defenses of their respective hosts in order to produce a successful infection. The ability to undergo meiosis may provide a survival benefit for these fungi by promoting successful infection. A characteristic central feature of meiosis is recombination between homologous chromosomes. This process is associated with repair of DNA damages, particularly double-strand breaks. The ability of *C. neoformans* and *U. maydis* to undergo meiosis may contribute to their virulence by removing the oxidative DNA damages caused by their host's release of reactive oxygen species.

Variations in Lifecycles

Many variations occur. Some are self-compatible and spontaneously form dikaryons without a separate compatible thallus being involved. These fungi are said to be homothallic, versus the normal heterothallic species with mating types. Others are secondarily homothallic, in that two compatible nuclei following meiosis migrate into each basidiospore, which is then dispersed as a pre-existing dikaryon. Often such species form only two spores per basidium, but that too varies. Following meiosis, mitotic divisions can occur in the basidium. Multiple numbers of basidiospores can result, including odd numbers via degeneration of nuclei, or pairing up of nuclei, or lack of migration of nuclei. For example, the chanterelle genus *Craterellus* often has six-spored basidia, while some corticioid *Sistotrema* species can have two-, four-, six-, or eight-spored basidia, and the cultivated button mushroom, *Agaricus bisporus*. can have one-, two-, three- or four-spored basidia under some circumstances. Occasionally, monokaryons of some taxa can form morphologically fully formed basidiomes and anatomically correct basidia and ballistic basidiospores in the absence of dikaryon formation, diploid nuclei, and meiosis. A rare few number of taxa have extended diploid lifecycles, but can be common species. Examples exist in the mushroom genera *Armillaria* and *Xerula*, both in the Physalacriaceae. Occasionally, basidiospores are not formed and parts of the "basidia" act as the dispersal agents, e.g. the peculiar mycoparasitic jelly fungus, *Tetragoniomyces* or the entire "basidium" acts as a "spore", e.g. in some false puffballs (*Scleroderma*). In the human pathogenic genus *Cryptococcus*, four nuclei following meiosis remain in the basidium, but continually divide mitotically, each nucleus migrating into synchronously forming nonballistic basidiospores that are then pushed upwards by another set forming below them, resulting in four parallel chains of dry "basidiospores".

Other variations occur, some as standard lifecycles (that themselves have variations within variations) within specific orders.

Rusts

Rusts (Pucciniales, previously known as Uredinales) at their greatest complexity, produce five different types of spores on two different host plants in two unrelated host families. Such rusts are heteroecious (requiring two hosts) and macrocyclic (producing all five spores types). Wheat stem rust is an example. By convention, the stages and spore states are numbered by Roman numerals. Typically, basidiospores infect host one, also known as the alternate or sexual host, the mycelium forms pycnidia, which are miniature, flask-shaped, hollow, submicroscopic bodies embedded in host tissue (such as a leaf). This stage, numbered "0", produces single-celled spores that ooze out in a sweet liquid and that act as nonmotile spermatia, and also protruding receptive hyphae. Insects and probably other vectors such as rain carry the spermatia from spermagonium to spermagonium, cross inoculating the mating types. Neither thallus is male or female. Once crossed, the dikaryons are established and a second spore stage is formed, numbered "I" and called aecia, which form dikaryotic aeciospores in dry chains in inverted cup-shaped bodies embedded in host tissue. These aeciospores then infect the second host, known as the primary or asexual host (in macrocyclic rusts). On the primary host a repeating spore stage is formed, numbered "II", the urediospores in dry pustules called uredinia. Urediospores are dikaryotic and can infect the same host that produced them. They repeatedly infect this host over the growing season. At the end of the season, a fourth spore type, the teliospore, is formed. It is thicker-walled and serves to overwinter or to survive other harsh conditions. It does not continue the infection process, rather it remains dormant for a period and then germinates to form basidia (stage "IV"), sometimes called a promycelium. In the Pucciniales, the basidia are cylindrical and become 3-septate after meiosis, with each of the 4 cells bearing one basidiospore each. The basidiospores disperse and start the infection process on host 1 again. Autoecious rusts complete their life-cycles on one host instead of two, and microcyclic rusts cut out one or more stages.

Smuts

The characteristic part of the life-cycle of smuts is the thick-walled, often darkly pigmented, ornate, teliospore that serves to survive harsh conditions such as overwintering and also serves to help disperse the fungus as dry diaspores. The teliospores are initially dikaryotic but become diploid via karyogamy. Meiosis takes place at the time of germination. A promycelium is formed that consists of a short hypha (equated to a basidium).

In some smuts such as *Ustilago maydis* the nuclei migrate into the promycelium that becomes septate (i.e., divided into cellular compartments separated by cell walls called *septa*), and haploid yeast-like conidia/basidiospores sometimes called sporidia, bud off laterally from each cell. In various smuts, the yeast phase may proliferate, or they may fuse, or they may infect plant tissue and become hyphal. In other smuts, such as *Tilletia caries*, the elongated haploid basidiospores form apically, often in compatible pairs that fuse centrally resulting in "H"-shaped diaspores which are by then dikaryotic. Dikaryotic conidia may then form. Eventually the host is infected by infectious hyphae. Teliospores form in host tissue. Many variations on these general themes occur.

Smuts with both a yeast phase and an infectious hyphal state are examples of dimorphic Basidiomycota. In plant parasitic taxa, the saprotrophic phase is normally the yeast while the infectious stage is hyphal. However, there are examples of animal and human parasites where the species are dimorphic but it is the yeast-like state that is infectious. The genus *Filobasidiella* forms basidia on hyphae but the main infectious stage is more commonly known by the anamorphic yeast name *Cryptococcus*, e.g. *Cryptococcus neo-formans* and *Cryptococcus gattii*.

The dimorphic Basidiomycota with yeast stages and the pleiomorphic rusts are examples of fungi with anamorphs, which are the asexual stages. Some Basidiomycota are only known as anamorphs. Many are yeasts, collectively called basidiomycetous yeasts to differentiate them from ascomycetous yeasts in the Ascomycota. Aside from yeast anamorphs, and uredinia, aecia and pycnidia, some Basidiomycota form other distinctive anamorphs as parts of their life-cycles. Examples are *Collybia tuberosa* with its apple-seed-shaped and coloured sclerotium, *Dendrocollybia racemosa* with its sclerotium and its *Tilachlidiopsis racemosa* conidia, *Armillaria* with their rhizomorphs, *Hohenbuehelia* with their *Nematoctonus* nematode infectious, state and the coffee leaf parasite, *Mycena citricolor* and its *Decapitatus flavidus* propagules called gemmae.

Development of Basidiomycetes

In a large number of Basidiomycetes the dikaryotic somatic mycelia, both secondary and tertiary intertwine producing highly organized sporocarps or fruiting bodies or fructifications or sporophores which are known as basidiocarps. Whereas, in a relatively few Basidiomycetes there is total absence of the development of basidiocarps.

The basidiocarps may be epigeous or hypogeous, or there may be many transitional forms between these conditions.

They are extremely variable in size, texture, colour and shape ranging from very large macroscopic to even minute microscopic in size. The basidiocarps besides being thin crust-like cartilaginous, or fleshy, spongy, gelatinous, corky, woody, may be almost of any other texture. So also in colour they may be bright orange-red, yellow, brown, dark, white and different grades of these and some other shades of colour.

The basidiocarps are often so highly developed and attract attention so easily that the main body of the fungus which is the extensive mycelium, usually goes unnoticed. Most common ones are crust-like, partly or completely shelved without any stalk; or stalked umbrella-like, fan-shaped, coralloid, round to oval and of various other configurations.

Structurally the basidiocarps may also vary enormously. But In general they are composed of fertile tissue, the hymenium or the hymenial layer produced by the secondary mycelium and sterile tissue, the trama or context of tertiary mycelium. The hymenial layer may be smooth or spread along plate-like structures known as gills or lamellae.

The part of the basidiocarp bearing the hymenium is known as hymenophore. The basidiocarp is often covered by a covering known as peridium.

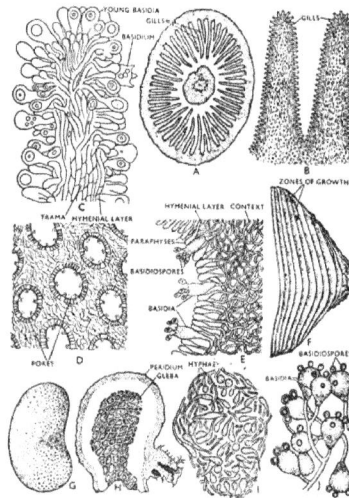

Structural details of basidiocarps. A. Complete section across the pileus of a basidiocarp of *Coprinus* sp. showing the radiating gills covered with blackened hymenium. B. Closer view of the gills. C. Details of a gill showing trama and hymenial layer. D. Portion of a section through a basidiocarp of *Polyporus* sp. showing pores lined with hymenium. E. Details of a portion of a pore showing hymenial layer composed of basidia bearing basidiospores and paraphyses, and context. F. Zones of growth of hymenial layers in the basidiocarp of *Fomes* sp. G. *Scleroderma* sp. An entire basidiocarp. H. Sectional view of a basidiocarp of *Scleroderma* sp. I. Immature gleba. J. Basidia with asterigmate basidiospore.

The Basidiocarp may be:

(i)   Open without being covered by a peridium from the very beginning of its development exposing the hymenium— gymnocarpous.

(ii) The hymenium is enclosed by basidiocarp tissue at first but later becomes exposed to the open air before the spores are mature—hemiangiocarpous.

(iii) The fertile portion (gleba) of the basidiocarp is enclosed in basidiocarp tissue until well after the spores are matured—angiocarpous, and

(iv) When the mature hymenium remains covered—endocarpous.

The hymenium may be borne on all sides of the basidiocarp—amphigenous, or on one side only—unilateral. In cases where the basidiocarps are closed or closed up to maturity of spores, enclosed within the peridium are the tramal plates lined with hymenium, known as gleba. The tramal plates and the hymenium are ultimately transformed into a characteristic mass, it is called glebal mass.

Where the basidiocarps are open from the very beginning of their development, the spores are dispersed mostly by a special mechanism known as fluid drop or drop-excretion mechanism of dispersal of spores. Whereas, in basidiocarps opening at a later stage, the spores are dispersed by agencies like wind, raindrops, animals and insects.

Again in others, where the basidiocarps remain closed, the spores are liberated by the disintegration or accidental breaking of the basidiocarps.

The hymenium of a basidiocarp consists of basidia and sterile structures interspersed with them. These sterile structures are often regarded as immature basidia. But they are generally designated as paraphyses (sing, paraphysis) and have been compared with those of the Ascomycetes. This has been questioned by Leutz (1954) and others.

According to them, the sterile structures of the hymenium are nothing but the sterile hyphal ends which are basically, just like basidia, nothing but the components of the dikaryotic secondary mycelium.

Whereas, the asci and paraphyses in the Ascomycetes originate from different hyphal elements. As such the similarities between the paraphyses of the Basidiomycetes and the Ascomycetes are superficial and they are not homologous structures. Hedwig (1789) was the first person to use the term 'paraphysis' in relation to fungi. Functionally the paraphyses are considered as spacing organs.

Besides paraphyses, the hymenium may also have certain other sterile structures whose taxonomic interest has been emphasized largely by Bourdot and Galzin (1928), and Overholts (1929).

Some of them are:

i. Acanthophyses (sing, acanthophysis):

These are sterile hyemenial hyphal ends having a number of short pin-like outgrowths on their surface. Burt (1914) designated them as 'bottle-brush paraphyses', present in certain species of Aleurodiscus.

ii. Dendrophyses (sing, dendrophysis):

These are hyphal ends of the hymenium or context having irregular tree-like branching. They are thick- or thin-walled and covered throughout with spines of variable length, also encountered in certain species of Aleurodiscus.

iii. Dichophyses (sing, dichophysis):

These sterile hyphal ends arising from the hymenium or subhymenium are antler-like in appearance because of their successively dichotomous, wide-angled branching and the prong-like terminal branchlets. They are present in some tropical species of Hymenochaete and species of Vararia.

iv. Pseudophyses (sing, pseudophysis):

These are smooth thin-walled, unbranched hymenial structures which are knobbed, knotted or nodose or moniliform. They are present in species of Aleurodiscus and Coprinus.

v. Gystidia (sing, cystidium):

These are hymenial or subhymenial organs, often projecting beyond basidia. They are elongated subcylindrical globose to sub- globose; hyaline to uniformly coloured or wall coloured and sap not coloured, never dark-brown or black even in KOH. They are present in species of Stereum, Pork, Armillariella, Coprinus, Lentinus, etc.

The cystidium has more or less rounded tip which may be smooth or provided with spines. The wall of the cystidium may have deposition of Ca-oxalate crystals, found in species of Peniophora.

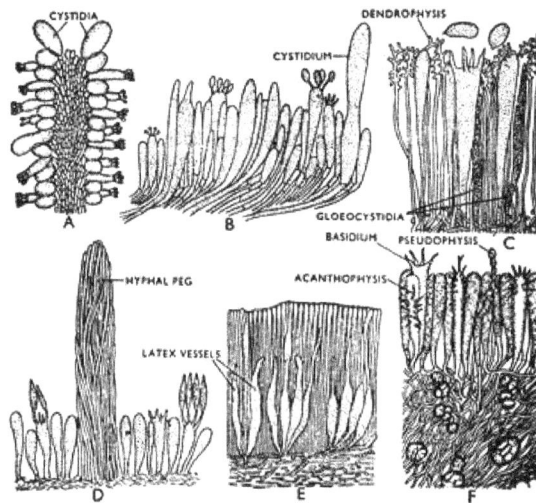

Various types of sterile structures of the hymenial layer. A. Cystidia in *Armillariella mellea*. B. Cystidium in *Stereum purpureum*. C. Dendrophysis and gloeocystidia in *Aleurodiscus sparsus*. D. Hyphal peg in *Epithele* sp. E. Latex vessels in *Corticium seriale*. F. Acanthophysis and pseudophysis in *Aleurodiscus oakesii*.

The existence of cystidia was first demonstrated by Micheli (1729) and was put to wider use by Hedwig (1789). The taxonomic importance of cystidia in the study of the family Thelephoraceae has been introduced by Massee (1889) and later on adopted by Burt (1914-1926) and many others. In some fungi, e.g., Vararia sp., cystidioid prolongations occur in the hymenium.

These structures do not possess staining qualities of true cystidia and are known as pseudocystidia (sing, pseudocystidium).

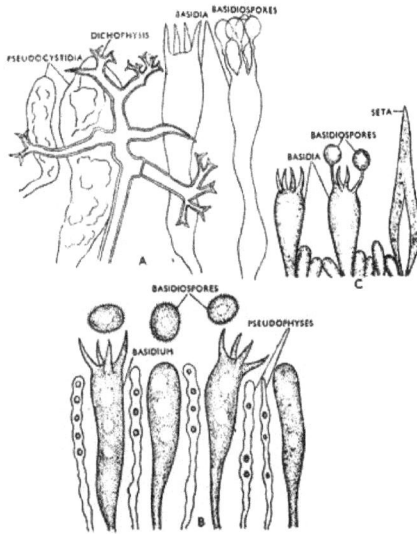

Various types of sterile structures of the hymenial layer. A. Dichophysis and pseudocystidia in *Vararia* sp. B. Pseudophyses in *Aleurodiscus amorphus*. C. Seta in *Fomes pini*.

vi. Gloeocystidia (sing, gloeocystidium):

These are subhymenial organs extending up to the hyemnium, may or may not remain associated with cystidia. They are of varied form, some of them being nearly indistinguishable from submerged, non-incrusted cystidia, others being long, flexuous, refractile structures.

A typical gloeocystidium is an elongated, usually rather slender, tortuous object, sometimes slightly enlarged toward the base and with a long slender apical neck, sometimes slender over the entire length and then often tapering very gradually toward the apex.

The walls are thin and colourless, very rarely with true septa. The walls of gloeocystidia are never incrusted with crystalline particles. The gloeocystidium contains sap which on mounting in glycerine divides into several opaque blocks, orange or brown in colour containing oily substances which do not exude when the gloeocystidium is broken.

The gloeocystidia are especially prominent in the genera Aleurodiscus, Stereum, and Merulius.

vii. Setae (sing, seta):

These are hymenial and subhymenial organs, mostly subhymenial and extended up to hymenium. The setae may be broadly ovate in shape and may extend beyond hymenial level. They are usually bristle or bristle-shaped bodies, typically deep yellow or brown in colour and dark-brown or black when treated with KOH. Setae contain brown-coloured sap.

Their tips may be pointed, blunt, curved or twisted. The genus Hymenochaete is characterized by having setae. Setae differ from most cystidia by their brown colour. When the setae have several radiating spines they are known as stellate setae. A stalked stellate seta is designated as asterophysis.

viii. Hyphal pegs:

Compound hyphal fasciculate projections derived from undifferentiated hyphae extend beyond the general level of the hymenium forming the hyphal pegs. Each hyphal peg consists of two or more parallel or interwoven hyphae forming a soft column. It may be pyramidal or cylindrical or conical in outline.

The hyphal pegs are found in the members of the families Thelephoraceae, Polyporaceae and Agaricaceae.

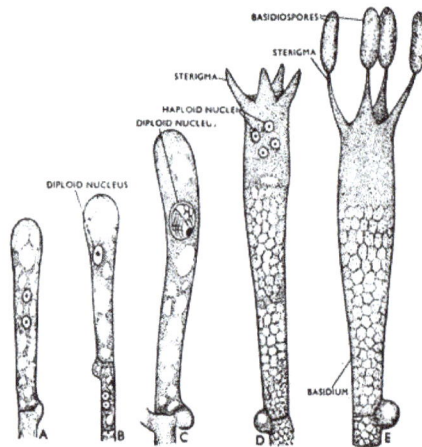

Successive stages in the development of a basidium and basidiospores. A. Dikaryotic terminal cell. B–C. Karyogamy. D. Young basidium with four haploid nuclei and four sterigmata. E. Mature basidium with four basidiospores borne on sterigmata.

ix. Latex vessels:

These are elongated, branched, non-septate, labyrinthiform, anastomosing hyphae of variable thickness. They rise from the subhymenium, penetrate the hymenium growing between the basidia and end on the upper surface of the hymenium. The latex vessels are multinucleate, vacuolate structures which contain milky or coloured or a hyaline sap which becomes coloured upon contact with air. They are present in Corticium seriate.

Development of Basidium and Basidiospores:

A basidium develops from a dikaryotic terminal cell of a hypha of the secondary mycelium. The dikaryotic terminal cell is separated from the rest of the hypha by a septum over which a clamp connection is usually found. During the development of the basidium the terminal cell enlarges to form young basidium, also known as basidiole.

The two nuclei of the basidiole fuse to form a diploid nucleus. The diploid nucleus then undergoes meiosis producing four haploid nuclei. The growing basidium then produces four protuberances at the top. These protuberances are known as sterigmata (sing, sterigma).

They may be lacking in some Basidiomycetes. When present, the tips of the sterigmata enlarge to produce basidiospore initials. The four haploid nuclei then squeeze through sterigmatal passage into each basidiospore initial. The basidiospore initial then becomes walled off to form a basidiospore with a single nucleus.

The portion of the basidiospore in contact with the sterigma has been designated by Heim (1931) as hilum; and hilar appendix or apiculus to the short, often sharp protrusion near the hilum at the basal end of the basidiospore.

Each basidium typically bears four basidiospores on four sterigmata, a tetrasterigmate basidium. The basidiospores may be borne symmetrically or asymmetrically on the sterigmata. Sometimes a basidium may produce only two sterigmata and two basidiospores. Such a basidium is called bisterigmate basidium.

In such case each of the two spores may receive two nuclei resulting in the development of binucleate basidiospores. In cases of bisterigmatic basidia the basidiospores may also be uninucleate. Here, the other two nuclei remain unused within the basidium and eventually disintegrate. The number of basidiospores may also be more than four.

The basidia are usually slender, clavate to broadly clavate. Spherical to elongate basidia are also not uncommon. They may be septate or aseptate. Septation is usually transverse or vertical, may also be oblique. During nuclear division the orientation of the nuclear spindles may also vary in different basidia.

When the nuclear spindles are oriented transversely to the basidium, such basidia are regarded as of the chiastobasidial type. Again basidia having nuclear spindles oriented longitudinally or obliquely are the stichobasidial type.

Following are the Main Types of basidia:

i. Homobasidia:

The basidia are slender, clavate to broadly clavate, or nearly globose and aseptate. The number of basidiospores may be four or more, with or without sterigmata. They are also known as autobasidia or holobasidia. The homobasidia may be chiasto-, or stichobasidial type.

When the basidia are developed inside a fruit body, they are the endobasidia. Again basidia with basidiospores symmetrical on sterigmata are the apobasidia and those with basidiospores asymmetrical are the autobasidia.

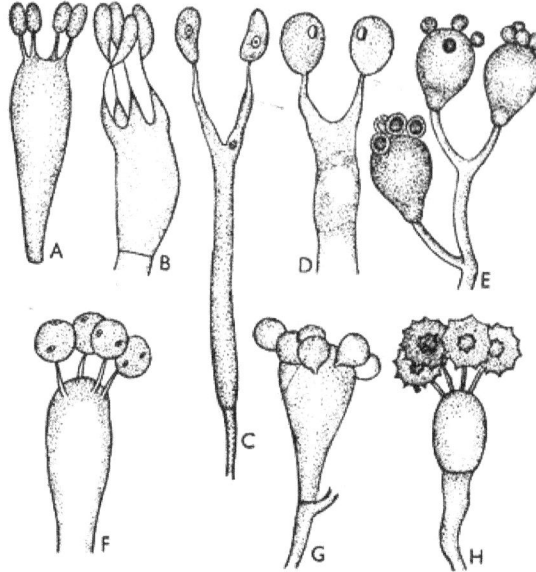

Various types of basidia. A, B, E and F–H. Homobasidia. C and D. Deeply-lobed heterobasidia—tuning fork type of basidia. A, B, F and H. Homobasidia bearing sterigmate basidiospores.
E and G. Homobasidia bearing asterigmate basidiospores.

ii. Heterobasidia:

The basidia are either septate or deeply lobed. They are also known as phragmobasidia or metabasidia.

In heterobasidia, the enlarged basal portion of the mature basidium in which nuclear fusion takes place has been designated by Neuhoff (1924) as hypobasidium and which bears upon it the epibasidium, but not the sterigmata directly. According to Neuhoff the hypobasidium and the epibasidium make up a heterobasidium.

Linder (1940) discarded this terminology and preferred probasidium for hypobasidium and basidium for epibasidium.

The Heterobasidia may again be:

  (i)   Transversely septate.

  (ii) Longitudinally septate, and

  (iii) Deeply lobed.

The transversely Septate Basidia may be differentiated into:

(a) With thin-walled hypobasidium and the epibasidium is traversed by three septa. Each cell of the epibasidium produces a long tube at whose apex a sterigmatic structure is formed on which arises the one-celled uninucleate basidiospore. The basidiospores are at the same level, as in the genus Auricularia.

(ii) The hypobasidium produces hooked epibasidium which again produces rather long sterigmata, seen in the genus Helicobasidium.

(ii) The hypobasidium is firm-walled from which grows out the straight or curved epibasidium which becomes four-celled. The basidiospores are borne on well-developed sterigmata, found in the genus Septobasidium.

(d) In the Uredinales and Ustilaginales the mature teleutospore, a uninucleate but diploid thick-walled structure, is the probasidium. On germination it gives rise to a tubular basidium, designated as a promycelium. By meiotic division of the diploid nucleus and by transverse septation, the promycelium becomes four-celled.

Each cell of the promycelium contains a single haploid nucleus which takes part in the development of a basidiospore. In the Uredinales the basidiospores are borne on sterigmata and number of basidiospores in four. Whereas, in the Ustilaginales there is absence of sterigmata and more than four basidiospores are formed.

The longitudinally septate basidia are the so-called cruciate type. The hypobasidium is longitudinally septate. The nuclear spindle is chiastobasidially oriented.

Two vertical or oblique septa are formed at right angles to one another producing a four-celled structure from each of which arises a prolongation—the epibasidium which is terminated by a sterigma, bearing the basidiospore. The various genera of the Tremellaceae exhibit this form a basidial character.

In the members of the Tulasnellaceae the basidium is not divided by vertical septa. The hypobasidium is subglobose, pyriform or broadly clavate. From the upper portion of the hypobasidium arise usually four stout cells—the epibasidia, each of which is narrowed at the tip to form a sterigma upon which a single basidiospore is formed. The epibasidia are separated from the hypobasidium by a septum at the base of each.

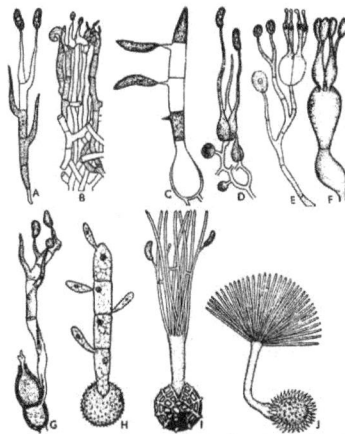

Various types of basidia. A–C. Transversely septate heterobasidia. A. Thin-walled hypobasidium in *Auricularia* sp. B. Hooked epibasidium in *Helicobasidium* sp. C. Firm-walled hypobasidium in *Septobasidium* sp. D–E. Longitudinal to obliquely septate heterobasidia. F. Heterobasidium with four epibasidia separated from hypobasidium by a septum at the base of each. G–J. Thick-walled probasidia with tubular basidia producing basidiospores. G. Basidiospores borne on sterigmata in *Puccinia* sp. H–J. Basidiospores not borne on sterigmata. H. *Ustilago* sp. I. *Tilletia* sp. J. *Neovossia* sp.

In the Dacrymycetaceae the basidia, when young, are long cylindrical or somewhat clavate. The basidial nucleus divides stichobasidially. The broadened apex of the basidium becomes deeply lobed to produce two epibasidia and almost of the same length as the hypobasidium. Each epibasidium is terminated by a sterigma at whose apex the basidiospore is produced.

The basidium takes a 'tuning fork' in appearance, from which it is known as tuning fork type of basidium.

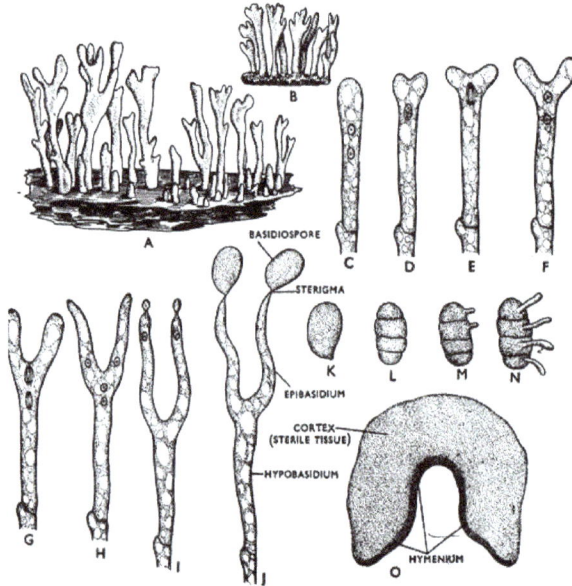

Dacryopinax sp. (*Guepinia* sp.) A and B. Basidiocarps at different stages of maturity attached to substratum. C to I. Successive stages in the development of a basidium and basidiospores. C. Dikaryotic terminal cell. D. Karyogamy. E to I. Stages of nuclear division, formation of epibasidium and hypobasidium, sterigmata and basidiospores. J. A mature basidium bearing basidiospores. K to N. Stages of basidiospore germination. O. Sectional view of a basidiocarp showing cortex (sterile tissue) and hymenium.

The Basidiospores:

The basidiospore is a haploid structure. Basidiospores are generally unicellular, uninucleate and thin-walled structures with the exception of some septate and some thick-walled ones. They may be globose oval, elongated, or sausage-shaped or apically truncated with cut top. Their colour en masse, may be hyaline, pink yellow, green, orange, brown, or black.

The basidiospore wall is almost universally smooth, except a few cases where wall markings are present. Some basidiospores give a positive amyloid reaction by being coloured blue with Melzer's reagent. Again some thick-walled spores absorb aniline blue dye strongly to the, inner-wall surface.

Such spores are said to be cyanophilous. Basidiospores are asterigmate when the sterigmata are lacking. But usually they are sterigmate, i.e., borne on sterigmata. The sterigmate basidiospores may be symmetrical or asymmetrical depending on whether they are arranged symmetrically or asymmetrically on the sterigmata. In majority of the Basidiomycetes, the basidiospores are arranged asymmetrically on the sterigmata.

In most of the Basidiomycetes where the basidiospores are produced in the open air there is usually a special provision for the discharge of these spores from their points of attachment. The spores are usually discharged with considerable violence. The term ballistospore has been applied to violently projected spores (Derx, 1948).

Whilst most basidiospores are ballistospores, some, e.g., those of the Gasteromycetes, are not. Basidiospores on a four-spored basidium are discharged in a regular succession and not simultaneously. The interval between discharge of the first and second spore may be a minute or more; that between discharge of the second and third, or the third and fourth spore may be somewhat longer.

Since the maturity of basidia is usually a prolonged process, the period of basidiospore discharge is also prolonged which may last for hours, days or even weeks. The present-day knowledge of spore discharge among the Basidiomycetes is from the researches of Buller (1909-1933).

Buller observed that in cases where the basidiospores are attached asymmetrically on the sterigmata, the spores are discharged violently by the mechanism which is termed the drop-excretion mechanism, also known as fluid drop mechanism.

A minute or so before the spore discharge, a small droplet of liquid exudes at the hilum (apiculus), it is often referred to as 'Buller's drop'. Within 5 to 20 seconds this liquid drop increases in volume to about one-fifth the size of the spore and then both spore and liquid drop suddenly shot off from the sterigma. The sterigma may or may not collapse after the spore is discharged.

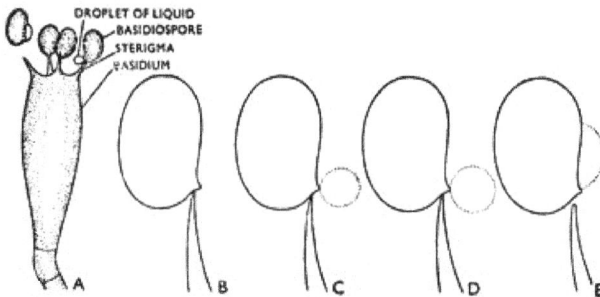

Discharge of basidiospores by drop-execretion mechanism. A. Basidiospore discharge. B-E, Successive stages in the fluid drop excretion and basidiospore discharge.

Buller (1909) initially postulated that basidiospore discharge was due to the sudden rounding-off of the contacting walls of the spore and sterigma. He subsequently (1922) rejected this hypothesis and suggested that a pressure developing in the vicinity of the hilum was the source of the force for discharge. Buller also suggested that the force for discharge is caused by surface tension energy.

Ingold (1939) explained that by the time the liquid drop flow is at the hilum, the spore also will tend to move in the opposite direction exerting pressure on the end of the sterigma. This pressure causes the discharge of spore. Corner (1948) stated that the liquid drop formed at the hilar appendix was surrounded by a membrane, which, he suggested, was an extension of the wall of the sterigma.

Olive (1964) suggested that the discharge was due to the explosion of a gas bubble in the vicinity of the hilar appendix and to the residual gas between the inner wall and the outer membrane of the spore and sterigma.

Wells (1965) with the electron microscope studied the ultras structure of the basidia and basidiospores in Schizophyllum commune. He observed that a thin fragile extension of the wall of the sterigma is present around the base of the basidiospore. The liquid drop is formed within that portion of the extension of the sterigma wall covering the hilar appendix.

The enlargement of the drop contributes to the rupturing of the contacting surfaces between the sterigma and the spore, and to the force of discharge. Besides this, the turgour pressure within the basidium also contributes to the force of basidiospore discharge.

The basidiospores germinate in presence of moisture by germ tubes and produce primary mycelium. The uninucleate basidiospores produce monokaryotic primary mycelium. But in case of binucleate basidiospores the condition is different. A basidiospore may be uninucleate in origin but subsequently become binucleate by the mitotic division of the single nucleus.

Such a binucleate basidiospore behaves like a uninucleate basidiospore. Whereas, it may so happen, two of the four haploid nuclei of the basidium may pass into a single basidiospore and the other two in another basidiospore. Here the binucleate basidiospores possess two nuclei which are genetically different. Such binucleate basidiospores give rise to dikaryotic secondary mycelium.

Again basidiospores instead of germinating by germ tubes may bud out large number of conidia from which the mycelium is produced.

In the Basidiomycetes there are both homothallic and heterothallic species. Since the term heterothallic as originally applied refers to mycelia representing different sexes, the use of the terms heterothallic and homothallic in the Basidiomycetes should be avoided, if not otherwise qualified.

It is advisable to use self-compatible or self-fertile and self-incompatible or interfertile for homothallic and heterothallic respectively. The large majority of the Basidiomycetes are self-incompatible and rather few are self-compatible.

Among the self-incompatible Basidiomycetes, the compatibility may be governed by one allelomorphic pair of factors, Aa, located on different chromosomes, or by two pairs of allelomorphic factors Aa Bb located on different chromosomes and segregating independently.

The former condition is known as bipolarity, the corresponding species bipolar species; and the latter is designated as tetrapolarity, the corresponding species tetrapolar species.

The common explanation of these phenomena is that in bipolar species fruit body development is dependent on two factors and in tetrapolar species on four. Basidium of a tetrapolar species again will produce two kinds or four kinds of basidiospores depending on the method of chromosome segregation during meiosis and on crossing-over.

When segregation of the factors controlling compatibility takes place in the first division, only two kinds of nuclei will result; AB, AB and ab, ab or Ab, Ab and aB, aB.

In such case when mycelium bearing nuclei containing AB or Ab unites with mycelium having nuclei possessing ab or aB, a dikaryotic condition, bearing all factors Aa Bb in each cell of the dikaryotic mycelium will be established to produce basidia and basidiospores.

If, however, segregation takes place at the second division of meiosis, a basidium will give rise to four types of spore: AB, Ab, aB, ab. Basidia and basidiospores are produced only when the secondary mycelium has the combination of compatibility factors of Aa, Bb.

Basidium

Basidium, in fungi (kingdom Fungi), the organ in the members of the phylum Basidiomycota (*q.v.*) that bears sexually reproduced bodies called basidiospores. The basidium serves as the site of karyogamy and meiosis, functions by which sex cells fuse, exchange nuclear material, and divide to reproduce basidiospores.

Structure

Most basidiomycota have single celled basidia (holobasidia), but in some groups basidia can be multicellular (a phragmobasidia). For instance, rust fungi in the order *Puccinales* have four-celled phragmobasidia that are transversely septate; some jelly fungi in the order Tremellales have four-celled phragmobasidia that are cruciately septate. Sometimes the basidium (metabasidium) develops from a probasidium, which is a specialized cell which is not elongated like a typical hypha. The basidium may be stalked or sessile.

The basidium typically has the shape of a club, where it is widest at the base of the hemispherical dome at its apex, and its base is about half the width of the greatest apical diameter. Versions where the basidium is shorter and narrower at the base are called "obovoid", and occur in genera such as *Paullicorticium*, *Oliveonia*, and *Tulasnella*. Basidia with a broad base are often described as "barrel-shaped".

Mechanism of Basidiospore Discharge

In most basidiomycota, the basidiospores are ballistospores—they are forcibly discharged. The propulsive force is derived from a sudden change in the center of gravity of the discharged spore. Important factors in forcible discharge include Buller's drop, a droplet of fluid that can be observed to accumulate at the proximal tip (hilar appendage) of each basidiospore; the offset attachment of the spore to the subtending sterigma, and the presence of hygroscopic regions on the basidiospore surface.

Upon maturity of a basidiospore, sugars present in the cell wall begin to serve as condensation loci for water vapor in the air. Two separate regions of condensation are critical. At the pointed tip of the spore (the hilum) closest to the supporting basidium, Buller's drop accumulates as a large, almost spherical water droplet. At the same time, condensation occurs in thin film on the adaxial face of the spore. When these two bodies of water coalesce, the release of surface tension and the sudden change in the center of mass leads to sudden discharge of the basidiospore. Remarkably, Money (1998) has estimated the initial acceleration of the spore to be about 10,000g.

Successful basidiospore discharge can only occur when there is sufficient water vapor available to condense on the spore.

Evolutionary Loss of Forcible Discharge

Some basidiomycetes lack forcible discharge, although they still form basidiospores. In each of these groups, spore dispersal occurs through other discharge mechanisms. For example, members of the order Phallales (stinkhorns) rely on insect vectors for dispersal; the dry spores of the Lycoperdales (puffballs) and Sclerodermataceae (earth balls and kin) are dispersed when the basidiocarps are disturbed; and species of the Nidulariales (bird's nest fungi) use a splash cup mechanism. In these cases the basidiospore typically lacks a hilar appendage, and no forcible discharge occurs. Each example is thought to represent an independent evolutionary loss of the forcible discharge mechanism ancestral to all basidiomycetes.

Basidiospore

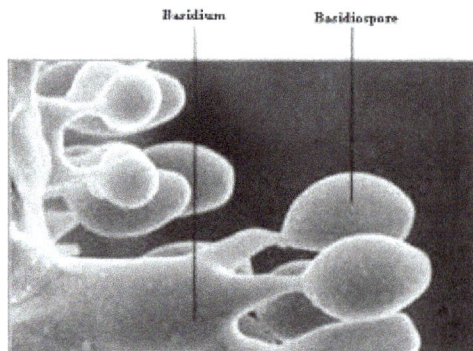

Basidiomycetes form sexual spores externally from a structure called a basidium. Four basidiospores develop on appendages from each basidium. These spores serve as the main air dispersal units for the fungi. The spores are released during periods of high humidity and generally have a night-time or pre-dawn peak concentration in the atmosphere.

General Structure and Shape:

Basidiospores are generally characterized by an attachment peg (called a hilar appendage) on its surface. This is where the spore was attached to the basidium. The hilar appendage is quite prominent in some basidiospore, but less evident in others. An apical germ pore may also be present. Many basidiospores have an asymmetric shape due to their development on the basidium. Basidiospores are typically single-celled (without septa), and typically range from spherical to oval to to oblong, to ellipsoid or cylindrical. The surface of the spore can be fairly smooth, or it can be ornamented.

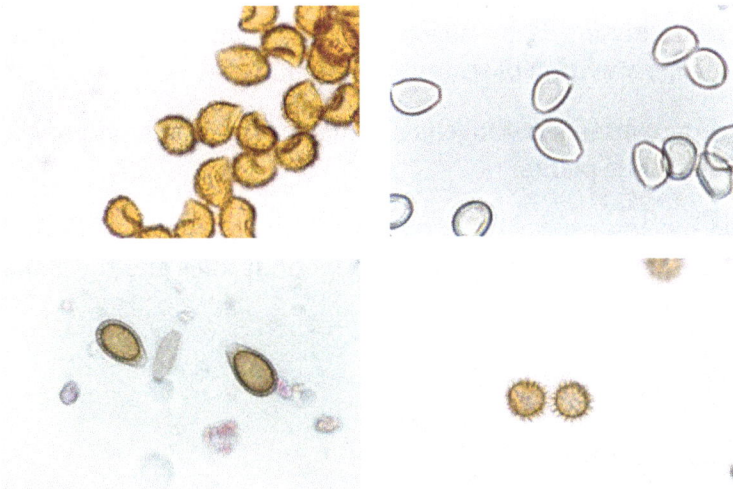

Blastocladiomycota

The Blastocladiomycota, a phylum of fungi that produce motile spores and gametes, live in water or in soils, where they are active when water is present and survive with thick-walled resting sporangia when soils are dry. Some genera are saprobic, but the phylum also contains pathogens of invertebrates and plants. Life cycles can include alternation of haploid and diploid generations and different generations may have different hosts. Sexual reproduction begins by fusion of gametes and meiosis takes place during germination of resistant sporangia. Saprobic members have been used as model organisms and genome sequencing of *Allomyces* and *Blastocladiella*continues this trend. *Coelomomyces* species parasitise mosquito larvae and are capable of causing death of up to 90% of their hosts. Obligate parasitism, narrow host ranges, need for an alternate host and unpredictable rate of control have deterred development of *Coelomomyces* species for mosquito control.

- Members of the Blastocladiomycota produce spores and gametes that are motile by means of a posteriorly directed whiplash flagellum.

- Motile spores, which are recognisable by their centrally located nucleus that is apically surrounded by a nuclear cap consisting of ribosomes, are recognisable at the ultrastructural level by microtubules in groups of three that extend from the apex of the kinetosome and surround the nucleus.

- Some saprobic species have an alternation of morphologically similar haploid and diploid generations and some pathogens have alternation of haploid and diploid generations with different morphologies and sometimes on different hosts.

- Sexual reproduction is by fusion of motile gametes and meiosis takes place during the germination of resting sporangia.

- The phylum contains pathogens of crustaceans, nematodes, algae and aquatic and semi-aquatic plants.

- Saprobic species, especially *Allomyces macrogynus* and *Blastocladiella emersonii*, are used for physiological and genetic studies and the genome of *A. macrogynus* has been sequenced.

Reproduction/Life Cycle

Sexual Reproduction

As stated above, some members of Blastocladiomycota exhibit alternation of generations. Members of this phylum also exhibit a form of sexual reproduction known as anisogamy. Anisogamy is the fusion of two sexual gametes that differ in morphology, usually size. In *Allomyces*, the thallus (body) is attached by rhizoids, and has an erect trunk on which reproductive organs are formed at the end of branches. During the haploid phase, the thallus forms male and female gametangia that release flagellated gametes. Gametes attract one another using pheromones and eventually fuse to form a Zygote. The germinated zygote produces a diploid thallus with two types of sporangia: thin-walled zoosporangia and thick walled resting spores (or sporangia). The thin walled sporangia release diploid zoospores. The resting spore serves as a means of enduring unfavorable conditions. When conditions are favorable again, meiosis occurs and haploid zoospores are released. These germinate and grow into haploid thalli that will produce "male" and "female" gametangia and gametes.

Asexual Reproduction

Similar to Chytridiomycota, members of Blastocladiomycota produce asexual zoospores to colonize new substrates. In some species, a curious phenomenon has been observed in the asexual zoospores. From time to time, asexual zoospores will pair up and exchange cytoplasm but not nuclei.

Ecological Roles

Plant leaf with *Physoderma menyanthis* (former *Cladochytrium menyanthis*) signs

Similar to Chytridiomycota, members of Blastocladiomycota are capable of growing on refractory materials, such as pollen, keratin, cellulose, and chitin. The best known species, however, are the parasites. Members of *Catenaria* are parasites of nematodes, midges, crustaceans, and even another blastoclad, *Coelomyces*. Members of the genus *Physoderma* and *Urophlyctis* are obligate plant parasites. Of economic importance is *Physoderma maydis*, a parasite of maize and the causal agent of brown spot disease. Also of importance are the species of *Urophlyctis* that parasitize alfalfa. However, ecologically, *Physoderma* are important parasites of many aquatic and marsh angiosperms. Also of human interest, for health reasons, are members of *Coelomyces*, an unusual parasite of mosquitoes that requires an alternate crustacean host (the same one parasitized by members of *Catenaria*) to complete its life cycle. Others that are ecologically interesting include a parasite of water bears and the zooplankter *Daphnia*.

Chytridiomycota

The Chytridiomycota is a phylum in the kingdom Fungi, whose members produce unwalled, asexual spores that swim by means of a single, posteriorly directed flagellum. Members are microscopic saprobes or parasites found in fresh and saline water and soils. The phylum contains two classes, the Monoblepharidomycetes with one order and the Chytridiomycetes with seven orders and several undescribed groups. Orders are based on molecular and ultrastructural characters. Many species of chytrids were described without these characters and need to be re-examined to place them in the proper orders and genera. Chytrids are important as degraders of cellulose, keratin and chitin and also as algal pathogens, sometimes controlling algal blooms. A few are plant pathogens with *Synchytrium endobioticum*causing black wart of potatoes. The rhizo-phydialean chytrid *Batrachochytrium dendrobatidis* grows in keratinised skin cells of amphibians and is pathogenic to many species, causing amphibian population declines on several continents and extirpation of some species.

- Members of the Chytridiomycota are ubiquitous in aquatic and terrestrial habitats and can be found by microscopically examining baits placed with aquatic debris or with soil plus water.

- Chytridiomycota reproduce primarily by zoospores, which are mitotically produced, contained by a membrane and motile via a posteriorly directed flagellum.

- Because of convergent evolution of light microscopic characters, many genera described before the molecular era are polyphyletic and may not be placed in the correct order.

- Systematics in the Chytridiomycota relies on transmission electron microscopic features of zoospores and on DNA sequence information.

- Algal parasites are usually species specific and are capable of affecting population levels of planktonic algae.

- *Batrachochytrium dendrobatidis* is a chytrid pathogen of amphibians responsible for some amphibian population declines on at least five continents.

Genome Structure

There are many different species within the classification of Chytridiomycota, and all have different genomes. For example, the species , there are eight mitochondrially-encoded tRNAs, and it is believed that they have at least one base pair mismatch at the first three positions of their aminoacyl acceptor stems. Bullerwell and Gray (2005) have developed a method of tRNA editing using the mitochondiral extract of *S. punctatus*. 5' tRNA editing occurs in the mitochondria of this species, as well as in the chytridiomycete order Monoblepharidales. However, it does not occur in *Rhizophydium brooksianum*. Laforest et. al. (2004) believe that 5' tRNA edition evolved twice independently within Chytridiomycota. The genome structure of *Batrachochytrium dendrobatidis* only five variable nucleotide positions, meaning it has a low level of genetic variation. Other characteristics of the *B. dendrobatidis* genome include nearly fixed heterozygous genotypes as well as chromosome lengtgh polymorphisms. The genes that encode alpha- and beta-tubulins are useful for encoding fungal phylogeny. Corradi et. al. (2004) studied the genetic similarities between members of Chrytidiomycota and arbuscular mycorrhizal fungi (AMF). They found large evolutionary differences at the amino-acid level, but at the mitochondiral level, differences were significantly smaller. Anaerobic chytridiomycetes contain an enzyme called pyruvate formate lyase, which is essential for f.ermentative formate production. This enzyme exists in the genetic structure of only one other eukaryotic lineage, chlorophytes.

Cell Structure and Metabolism

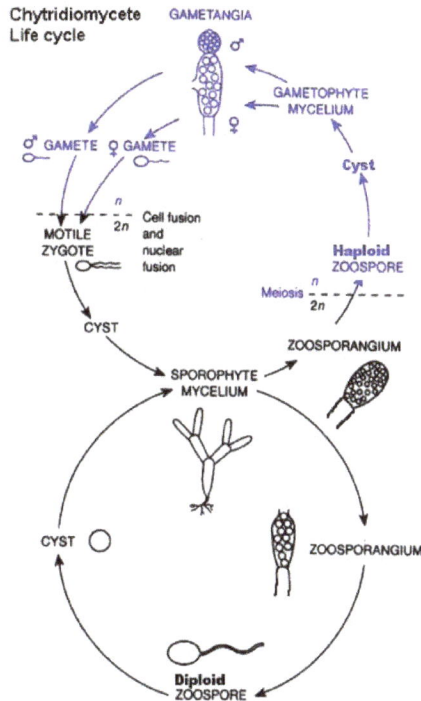

Diagram of life cycle. Microbiology by Joan Slonczewski

Chytridiomycota have unicellular or mycelial thalli. Cell walls are made of chitin, although one group has walls made of cellulose. Cell growth can be unicellular, or it can occur in the multicellular mycelium of aseptate hyphae. The thallus is typcially unicellular; it may also have limited hyphal growth. It is not considered mycelial. Hyphal cells are coenocytic, although this is not the case where there are reproductive structures. The ultrastructure of the zoospore is a definitve characteristic of Chytridiomycota. In the structure, ribosomes are aggregated around a nucule that is not enclosed in a nuclear cap. A nuclear cap is an extension of the nuclear membrane. Chytridiomycota have one or two flagella.

Chytridiomycota feed on both living and decaying organisms. They are heterotrophic.

Asexually, Chytridiomycota reproduce through the use of zoospores. In asexual reproduction, zoospores will swim until a desireable substrate is located. The zoospore attaches itself, feeds off its host; the cytoplasm grows, meiotic divisions occur, and a cell wall forms around the original zoospore. Protoplasm increases as the cell continues to develop. Finally, cleavage of the protoplasm occurs, which produces individual zoospores that are released through a pore. Sexual reproduction is haploid dominant. It also depends on the isomorphic alternation of generations. The haploid thallus, called the gametothallus, produces female and male gametes. These occur in pairs and are terminal and subterminal. Male gametes are orange-colored, while female gametes are colorless. In addition, female gametes are much larger than male gametes. Males are attracted to females when they produce the hormone sirenin, and females are attracted to males when they produce the hormone parisin.

The diploid thallus is called the sporothallus. The sporothallus produces two types of zoosporgia: zoosporgangium (meitosporangium) and resistant sporangium (meiosporangium). Zoosporangia produce diploid zoospores, which can function as a means of asexual reproduction. Sexual reproduction may be isogamous, anisogamous, or oogamous. One species, *Allomyces macrogynus*, has a sporic life cycle, something that occurs in plants but is rare to fungi. *A. macrogynus* is an example of an anisogamous species.

Life Cycle and Body Plan

Chytridiomycota are unusual among the Fungi in that they reproduce with zoospores. For most members of Chytridiomycetes, sexual reproduction is not known. Asexual reproduction occurs through the release of zoospores (presumably) derived through mitosis.

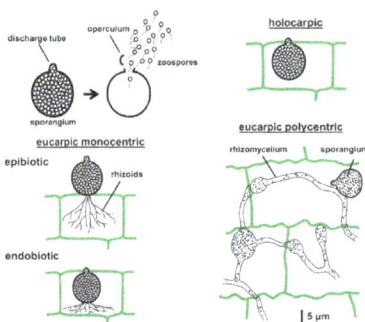

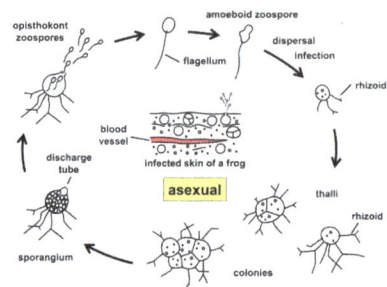

Types of chytrid thalli   Life cycle of *Batrachochytrium dendrobatidis*

Where it has been described, sexual reproduction of Chytridomycetes occurs via a variety of methods. It is generally accepted that the resulting zygote forms a resting spore, which functions as a means of surviving adverse conditions. In some members, sexual reproduction is achieved through the fusion of isogametes (gametes of the same size and shape). This group includes the notable plant pathogens *Synchytrium*. Some algal parasites practice oogamy: a motile male gamete attaches itself to a nonmotile structure containing the female gamete. In another group, two thalli produce tubes that fuse and allow the gametes to meet and fuse. In the last group, rhizoids of compatible strains meet and fuse. Both nuclei migrate out of the zoosporangium and into the conjoined rhizoids where they fuse. The resulting zygote germinates into a resting spore.

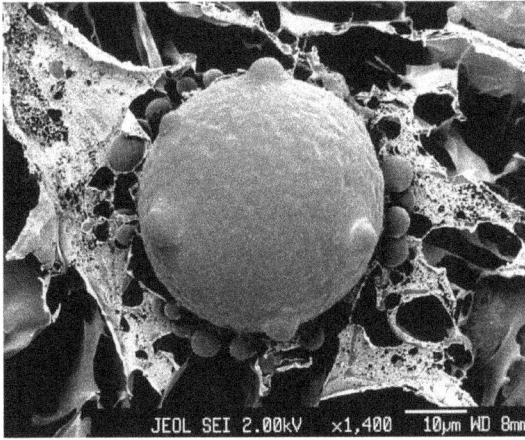

Sporangium and zoospores of the chytrid fungus *B. dendrobatidis*, under SEM

Sexual reproduction is common and well known among members of the Monblepharidomycetes. Typically, these chytrids practice a version of oogamy: the male is motile and the female is stationary. This is the first occurrence of oogamy in kingdom Fungi. Briefly, the monoblephs form oogonia, which give rise to eggs, and antheridia, which give rise to male gametes. Once fertilized, the zygote either becomes an encysted or motile oospore, which ultimately becomes a resting spore that will later germinate and give rise to new zoosporangia.

Upon release from the germinated resting spore, zoospores seek out a suitable substrate for growth using chemotaxis or phototaxis. Some species encyst and germinate directly upon the substrate; others encyst and germinate a short distance away. Once germinated, enzymes released from the zoospore begin to break down the substrate and utilize it produce a new thallus. Thalli are coenocytic and usually form no true mycelium (having rhizoids instead).

Chytrids have several different growth patterns. Some are holocarpic, which means they only produce a zoosporangium and zoospores. Others are eucarpic, meaning they produce other structures, such as rhizoids, in addition to the zoosporangium and zoospores. Some chytrids are monocentric, meaning a single zoospore gives rise to a single zoosporangium. Others are polycentric, meaning one zoospore gives rise to many zoosporangium connected by a rhizomycelium. Rhizoids do not have nuclei while a rhizomycelium can.

Growth continues until a new batch of zoospores are ready for release. Chytrids have a diverse set of release mechanisms that can be grouped into the broad categories of operculate or inoperculate. Operculate discharge involves the complete or incomplete detachment of a lid-like structure, called an operculum, allowing the zoospores out of the sporangium. Inoperculate chytrids release their zoospores through pores, slits, or papillae.

Habitats

Synchytrium endobioticum on potatoes

Chytrids are aquatic fungi, though those that thrive in the capillary network around soil particles are typically considered terrestrial. The zoospore is primarily a means of thoroughly exploring a small volume of water for a suitable substrate rather than a means of long range dispersal.

Chytrids have been isolated from a variety of aquatic habitats, including peats, bogs, rivers, ponds, springs, and ditches, and terrestrial habitats, such as acidic soils, alkaline soils, temperate forest soils, rainforest soils, arctic and Antarctic soils. This has led to the belief that many chytrid species are ubiquitous and cosmopolitan. However, recent taxonomic work has demonstrated that this ubiquitous and cosmopolitan morphospecies hide cryptic diversity at the genetic and ultrastructural levels. It was first thought aquatic chytrids (and other zoosporic fungi) were primarily active in fall, winter, and spring. However, recent molecular inventories of lakes during the summer indicate that chytrids are an active, diverse part of the eukaryotic microbial community.

One of the least expected terrestrial environments the chytrid thrive in are periglacial soils. The population of the Chytridiomycota species are able to be supported even though there is a lack of plant life in these frozen regions due to the large amounts of water in periglacial soil and pollen blowing up from below the timberline.

Ecological Functions

Batrachochytrium Dendrobatidis

Dead frog with chytridiomycosis (*B. dendrobatidis*) signs

The chytrid *Batrachochytrium dendrobatidis* is responsible for chytridiomycosis, a disease of amphibians. Discovered in 1998 in Australia and Panama this disease is known to kill amphibians in large numbers, and has been suggested as a principal cause for the worldwide amphibian decline. Outbreaks of the fungus were found responsible for killing much of the Kihansi Spray Toad population in its native habitat of Tanzania, as well as the extinction of the golden toad in 1989. Chytridiomycosis has also been implicated in the presumed extinction of the Southern Gastric Brooding Frog , last seen in the wild in 1981, and the Northern Gastric Brooding Frog, last recorded in the wild in March 1985 . The process leading to frog mortality is thought to be the loss of essential ions through pores made in the epidermal cells by the chytrid during its replication.

Recent research has revealed that elevating salt levels slightly may be able to cure chytridiomycosis in some Australian frog species although further experimentation is needed.

Other Parasites

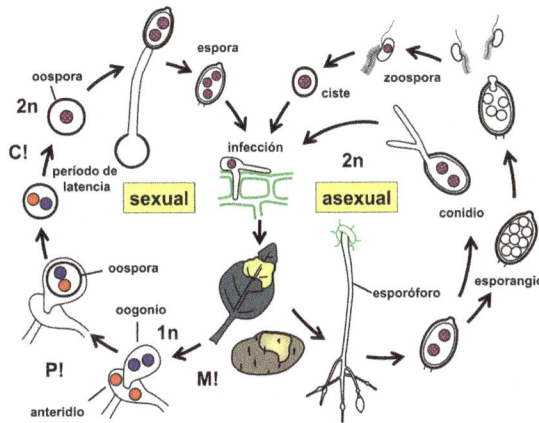

Life cycle of *Synchytrium endobioticum* in potato

Chytrids mainly infect algae and other eukaryotic and prokaryotic microbes. The infection can be so severe as to control primary production within the lake. It has been suggested that parasitic chytrids have a large effect on lake and pond food webs. Chytrids may also infect plant species; in particular, *Synchytrium endobioticum* is an important potato pathogen.

Saprobes

Arguably, the most important ecological function chytrids perform is decomposition. These ubiquitous and cosmopolitan organisms are responsible for decomposition of refractory materials, such as pollen, cellulose, chitin, and keratin. There are also chytrids that live and grow on pollen by attaching threadlike structures, called rhizoids, onto the pollen grains. This mostly occurs during asexual reproduction because the zoospores that become attached to the pollen continuously reproduce and form new chytrids that will attach to other pollen grains for nutrients. This colonization of pollen happens during the spring time when bodies of water accumulate pollen falling from trees and plants.

Glomeromycota

The Glomeromycota is a newly established phylum which comprises about 230 species that all live in close association with the roots of trees. Fossil records indicate that trees and their root symbionts share a long evolutionary history. It appears that all members of this family form arbuscular mycorrhizae: the hyphae interact with the root cells forming a mutually beneficial association where the plants supply the carbon source and energy in the form of carbohydrates to the fungus, and the fungus supplies essential minerals from the soil to the plant.

Reproduction and Life Cycle

There is no evidence that the Glomeromycota reproduce sexually. Studies using molecular marker genes have detected no genetic recombination or only low levels. Therefore it is generally assumed that the spores are formed asexually. There are conflicting reports on the question whether the nuclei in the mycelium and spores of one organism are genetically identical or not. The presence of multiple, slightly differing variants of the nuclear-encoded ribosomal RNA genes in single spores may or may not be due to this possible nuclear heterogeneity.

Under favorable conditions glomeromycotan spores germinate, form appressoria on host roots and establish a new mycorrhizal symbiosis. New spores may be formed on the mycelium either within or outside the root. In addition to propagation by spores, many species of Glomeromycota can colonize host plants from hyphal fragments in the soil or directly from symbionts that inhabit the roots of a neighboring plant.

Given that they are obligate symbionts, if no host root is found by the germinating hypha of a spore, growth ceases after some time, and the cytoplasm may be retracted within the spore. Because they cannot be cultured axenically, these fungi are propagated primarily on the host plant in pot cultures grown in a greenhouse. The spores produced in open pot cultures are not sterile and therefore harbor a wide variety of bacteria and other fungi. Cultures can be started using field soil containing spores or hyphae, from a number of purified spores that morphologically appear to represent a single species, or from a single spore. Glomeromycotan spores can be concentrated by wet-sieving methods or by centrifugation techniques. In addition to propagation via pot cultures, a number of glomeromycotan species can be grown in root organ cultures, i.e. on the roots of a plant growing on a sterile nutrient medium in a petri dish. The fungal biomass produced in root organ cultures usually does not contain other micro-organisms, and this is therefore the method of choice for certain molecular biological experiments.

Figure: Spores of *Scutellospora castanea*. Spore diameter approximately 220 μm.

Ecology and Physiology

Associations between plant roots and fungi (mycorrhizas) are ubiquitous. The great majority of land plants are host for some type of mycorrhiza. Members of most plant families form AM. Other types of mycorrhiza are formed with fungi from the phyla Asco- or Basidiomycota: ectomycorrhiza of trees and shrubs, ericoid mycorrhiza of Ericales, orchid mycorrhiza, and some others. Only a few plant families are regarded as non-mycorrhizal, among them the Brassicaceae, Caryophyllaceae and Chenopodiaceae.

As most crops are hosts for AM, this association is potentially an important resource for agriculture. Positive effects of the arbuscular mycorrhiza on plant growth, nutrient uptake and disease resistance have often been reported. These effects are relatively easy to demonstrate in comparison to sterilized soil which rarely is found under field conditions. Although the fungal diversity in temperate agricultural soils was shown to be low compared to natural field sites, there is almost always an indigenous fungal community the introduced AM fungi have to compete with.

AM has been reported to improve the uptake of different mineral nutrients. However, phosphate has been in the focus because it can be a limiting factor for plant growth due to its immobility in the soil. The fine hyphal network is superior to the relatively thick roots and root hairs in accessing phosphate in the soil. On the other hand, high available phosphate concentrations often seem to induce a limitiation of fungal colonization levels by the plants.

Host specificity of AM appears to be be very low because many species have been shown to colonize a wide range of host plants in the greenhouse. Interestingly, plants are typically colonized by a mixture of AM fungal species, often within the same root. Nevertheless, favorable and less favorable combinations of plant-fungal symbionts have been reported. AM fungi have also been shown to exert a specific influence on the species composition of plant communities. Field studies using molecular identification methods have demonstrated that distinct fungal communities are associated with different hosts.

Although the symbiosis is generally thought to be beneficial, under certain conditions the balance between the symbionts may be disturbed. There is evidence that a fungal symbiont may decrease plant growth. Conversely, certain non-photosynthetic plants may cheat the fungus by obtaining all their nutrients from them, including carbohydrates.

The cross-talk between symbionts and the genes involved in the establishment and maintenance of the symbiosis have been subject of intensive research efforts.

Although the mycorrhizal status of many species placed in this group has in fact not been demonstrated, only one fungus in the Glomeromycota is currently known which forms a different type of symbiosis: *Geosiphon pyriformis*. This fungus produces bladders that harbor symbiotic cyanobacteria. Nevertheless, molecular phylogenetic analysis has shown that *Geosiphon* is a member of the Glomeromycota.

Discussion of Phylogenetic Relationships

At the present time, the phylogeny presented here is based entirely on analyses of the small subunit RNA gene. Although additional genes have begun to be sequenced from some taxa phylogenetic hypotheses based on multilocus DNA sequence data have yet to be incorporated into their classification.

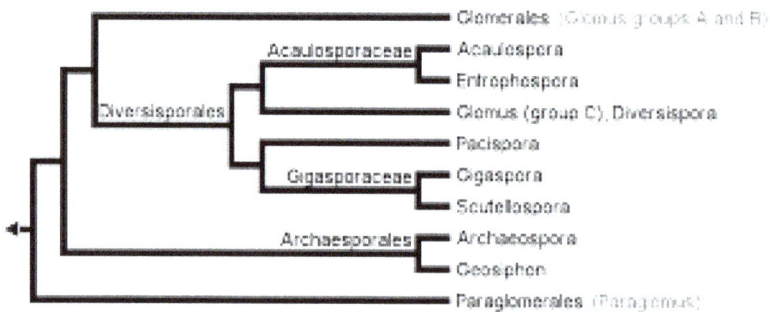

Phylogenetic tree of Glomeromycota lineages based on analyses of ribosomal small subunit sequences. *Glomus* subgroups as defined by Schwarzott et al.

rDNA phylogenies have shown that the genus *Glomus* is several times polyphyletic (Redecker et al., 2000b; Schwarzott et al., 2001). Species forming *Glomus*-like spores can be found in six different lineages within the Glomeromycota. The simple morphology of the spores apparently has concealed the large genetic variation within the polyphyletic genus as previously defined. *Paraglomus* appears to be the earliest-diverging glomeromycotan lineage in rDNA phylogenies, although sometimes receiving relatively weak bootstrap support. The separation of *Pacispora* and the *Diversispora* clade from other "*Glomus*" lineages" is well-supported by rDNA data.

Glomus groups A and B are exemplified by the well-known species *Glomus mosseae* and *Glomus claroideum*, respectively. The two groups are genetically relatively distant but still form a monophyletic group in rDNA phylogenetic trees.

The formation of a "sporiferous saccule" was once thought to be characteristic for the Acaulosporaceae (*Acaulospora* and *Entrophospora*), but now is known to occur in at least one additional lineage, namely *Archaeospora*. The Gigasporaceae (*Scutellospora* and *Gigaspora)* are distinguished by the formation of their spores on a "bulbous suspensor" and are well-supported by molecular data. Gigasporaceae and Acaulosporaceae form a clade in most rDNA phylogenies, which is in conflict with previous morphology-based analyses that placed *Glomus* and Acaulosporaceae together.

Relationships of Glomeromycota to other Fungi

The "Glomales" were previously placed in the Zygomycota, but the following evidence indicates that they form a monophyletic group distinct from other Zygomycotan lineages: their symbiotic habit, the apparent lack of zygospores and the rDNA phylogeny. Based on this evidence, Schüßler et al. (2001) erected the phylum Glomeromycota. These authors also corrected the previously used, grammatically incorrect ordinal name "Glomales" to "Glomerales". In phylogenetic trees based on rDNA, the Glomeromycota are the sister group to Asco- and Basidiomycota.

Classification

Since molecular phylogenetic methods have been used to elucidate the phylogenetic relationships among these fungi, their classification has been in a rapid transition. Molecular field studies have also revealed a large number of putative new species, suggesting that the 150 morphologically-defined species may vastly underestimate species diversity.

Traditionally, glomeromycotan taxonomy has been based on the morphology of the spores. The way the spore is formed on the hypha ("mode of spore formation") has been important to circumscribe genera and families, and the layer structure of the spore wall to distinguish species. Some species have a richly ornamented spore surface.

Glomeromycotan taxonomy is relatively young. Prior to 1974, most AM fungi were in the genus *Endogone*. However, in 1974 Gerdemann and Trappe (1974) removed the AM fungi from *Endogone* and placed them in four separate genera: *Glomus, Sclerocystis, Acaulospora* and *Gigaspora*. Unlike the putatively asexual members of the Glomeromycota, *Endogone* species reproduce sexually via zygospores, indicating their phylogenetic link with the phylum Zygomycota. Phylogenetic analysis of the nuclear small subunit ribosomal RNA strongly suggests that *Endogone* (Endogonales) and the Glomeromycota do not form a clade.

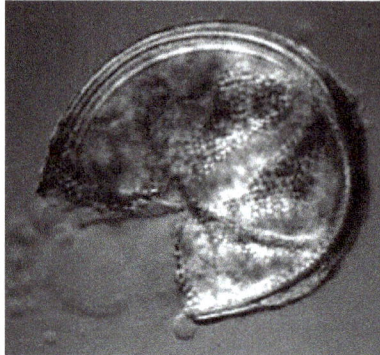

Spore of *Archaeospora leptoticha* (isolate NC176) cracked open under a microscope cover slide. The spore wall is multilayered with a richly ornamented surface. Uncracked spore diameter approximately 200 μm.

The order Glomales was erected by Morton and Benny (1990) to contain all AM fungi. At that time two more genera, *Scutellospora* and *Entrophospora*, had been established by other authors and three families (Glomaceae, Gigasporaceae and Acaulosporaceae) were recognized. These families were characterized by the mode of spore formation and were initially supported by molecular data.

Since then, it has become clear, however, that the mode of spore formation is not a useful diagnostic feature for some genera. Spores of the *Glomus* and *Acaulospora* types were reported to be produced by several distinct, deeply divergent lineages. Subsequently, they were described as two new genera *Archaeospora* and *Paraglomus* and placed in separate families. Because some species in *Archaeospora* were even dimorphic, members of this genus were classified originally in separate families.

Molecular phylogenetic analysis has also shown that species which form complex sporocarps formerly placed in the genus *Sclerocystis* are actually phylogenetically nested within well-characterized *Glomus* species with simple spores.

Section of a sporocarp of *Glomus sinuosum* (isolate MD126, formerly *Sclerocystis sinuosa*). Spores are arranged around a center of interwoven hyphae and covered by a "peridium". Sporocarp diameter approximately 250 μm.

A new genus *Pacispora* was erected by Oehl and Sieverding (2004), comprising some former *Glomus* species. The genus name *Gerdemannia*published for the same fungal group a few weeks later, is a synonym of *Pacispora*, and an illegitimate name based on the publication date. The spores of *Pacispora* have characteristics intermediate between *Glomus* and the Gigasporaceae.

Another emerging genus split off from *Glomus* is *Diversispora*. Only one *Glomus* species has been formally renamed so far, mainly based on ribosomal small subunit sequence signatures.

Zygomycota

The Zygomycota, or conjugation fungi, include molds, such as those that invade breads and other food products. The identifying characteristics of the Zygomycota are the formation of a zygospore during sexual reproduction and the lack of hyphal cell walls except in reproductive structures. Many (~100 species) are known plant root symbionts.

Structure

The mycelia of Zygomycota are divided into three types of hyphae. The rhizoids reach below the surface and function in food absorbtion. Above the surface, sporangiophores bear the spore-producing sporangia. Groups of rhizoids and sporangiophores are connected above the surface by stolons. Cell walls separating individual cells are absent in all but reproductive structures, allowing cytoplasm and even nuclei to move between cells.

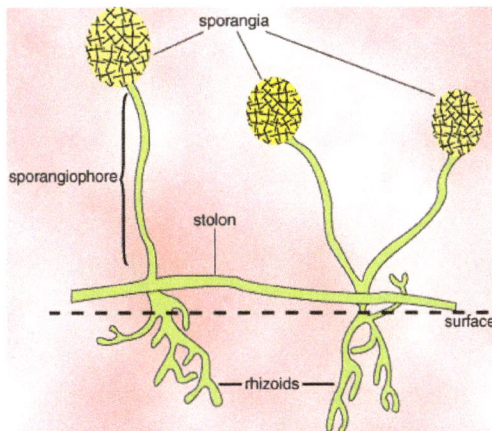

Structure of the Zygomycota

Characteristics

Zygomycota, like all true fungi, produce cell walls containing chitin. They grow primarily as mycelia, or filaments of long cells called hyphae. Unlike the so-called 'higher fungi' comprising the Ascomycota and Basidiomycota which produce regularly septate mycelia, most Zygomycota form hyphae which are generally coenocytic because they lack cross walls or septa. There are, however, several exceptions and septa may form at irregular intervals throughout the older parts of the mycelium or are regularly spaced in two sister orders of Zygomycota, the Kickxellales and Harpellales.

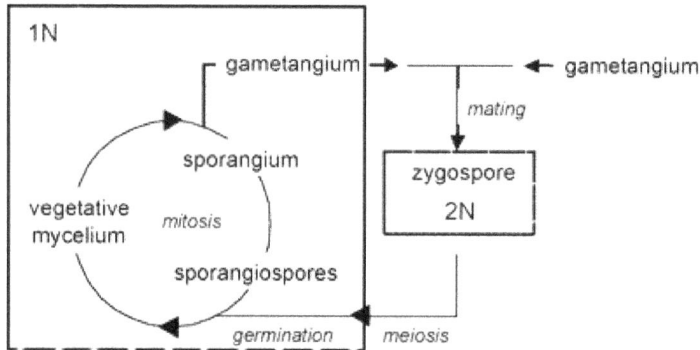

Generalized life cycle of Zygomycota. Asexual reproduction occurs primarily by sporangiospores produced by mitosis and cell division. The only diploid (2N) phase in the life cycle is the zygospore, produced through the conjugation of compatible gametangia during the sexual cycle

The unique character (synapomorphy) of the Zygomycota is the zygospore. Zygospores are formed within a zygosporangium after the fusion of specialized hyphae called gametangia during the sexual cycle. A single zygospore is formed per zygosporangium. Because of this one-to-one relationship, the terms are often used interchangably. The mature zygospore is often thick-walled, and undergoes an obligatory dormant period before germination. Most Zygomycota are thought to have a zygotic or haplontic life cycle. Thus, the only diploid phase takes place within the zygospore. Nuclei within the zygospore are believed to undergo meiosis during germination, but this has only been demonstrated genetically within the model eukaryote Phycomyces blakesleeanus.

Zygomycota typically undergo prolific asexual reproduction through the formation of sporangia and sporangiospores. Sporangiospores are distinguished from other types of asexual spores, such as conidia of the Ascomycota and Basidiomycota, by their development. Walled sporangiospores are formed by the internal cleavage of the sporangial cytoplasm. At maturity, the sporangial wall typically disintegrates or dehisces, thereby freeing the spores that are usually dispersed by wind or water.

Sporangia are formed at the ends of specialized hyphae called sporangiophores. In the model organism, Phycomyces blakesleeanus, sporangial development has been studied extensively to understand the genetic basis for various trophisms, including the strong phototrophic responses to blue light. A unique spore dispersal strategy for the Mucorales is exhibited by the dung fungus Pilobolus, whose name literally means 'the hat thrower'. The entire black sporangium is explosively shot off of the top of the sporangiophore up to distances of several meters. Phototrophic growth of the sporangiophore facilitates dispersal away from the dung onto a fresh blade of grass where it may be consumed by an herbivore, thereby completing the asexual cycle after the spores pass through the digestive system. Some members of the Entomophthorales (e.g., Basidiobolus, Conidiobolus) also reproduce via forcibly discharged asexual spores. Interestingly, species of Basidiobolus, Conidiobolus and several other genera produce a second kind of spore on a long stalk that appears to have certain morphological adaptations for efficient insect dispersal.

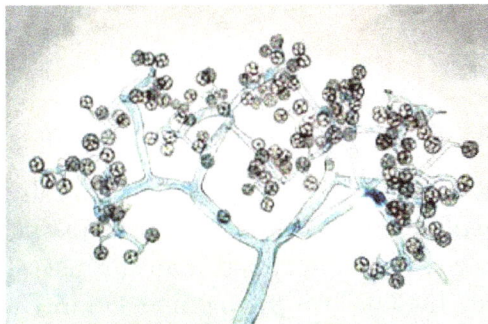

Dichotomously branching sporangiophore of Thamnidium elegans (Mucorales). The few-spored sporangiola are borne at the tips of the sporangiophore branches

Two variant types of sporangia include sporangiola and merosporangia. Sporangiola are simply uni-to-few spored sporangia containing between 1-to-30 spores. Merosporangia are elongated sporangiola with uniseriate spores usually produced from a vesicle or stalk.

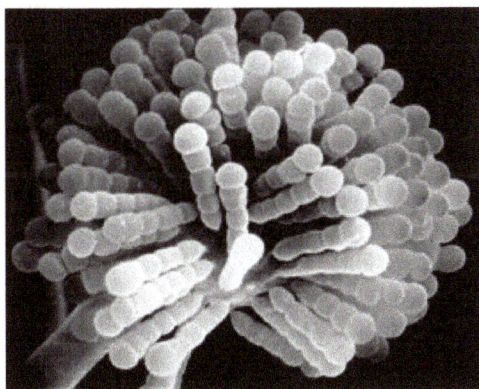

Scanning electron micrograph of uniseriate merosporangia produced on a vesicle (hidden beneath merosporangia) of Syncephalastrum racemosum (Mucorales)

Merosporangiferous members of the Zygomycota, however, do not form a clade, indicating that this sporangial type has evolved independently more than once within the phylum (e.g., Mucorales and Zoopagales). A unique sporangiolum type is the trichospore, a one-spored sporangiolum, produced by members of the Harpellales, which are endocommensals living within the gut of arthropods, including terrestrial beetles and millipedes, fiddler crabs, and the larvae of many aquatic insects. Trichospores possess one to several basal hair-like filaments that likely aid in the attachment of the spores to debris and plants in aquatic ecosystems before they reenter the arthropod gut.

Photo of thallus of Genistellospora homothallica (Harpellales) bearing trichospores attached to the hindgut cuticle of a Chilean blackfly.

Like other Fungi, Zygomycota are heterotrophic and typically grow inside their food, dissolving the substrate with extracellular enzymes, and taking up nutrients by absorption rather than by phagocytosis, as observed in many protists. The most common members of the Zygomycota are the fast growing members of the Mucorales. They function as decomposers in soil and dung, thereby playing a significant role in the carbon cycle.

Zygomycota also participate in a number of interesting symbioses. As mentioned above, the Harpellales inhabit arthropods (particularly freshwater aquatic insect larvae) where they are attached to the chitinous lining of the hindgut. Harpellids presumably feed on nutrients that are not utilized by the arthropod. Because they are generally assumed to neither harm nor benefit the host animal, this association is considered commensalistic. In contrast, the Entomophthorales include many insect pathogens that can cause huge disease outbreaks. Some of this pathogenicity is being tapped for use in the biocontrol of specific insect pests, including periodical cicadas. A number of other Zygomycota are mycoparasitic, or parasites of other fungi. All members of the Dimargaritales (only 15 species) and many Zoopagales are typically obligate parasites of mucoralean hosts. Other mycoparasites in the Mucorales specialize on mushroom fruiting bodies.

Sporangia of Spinellus fusiger (Mucorales) parasitic on fruitbodies of the mushroom Mycena pura

Certain species of Zoopagales parasitize non-fungal hosts, such as nematodes, rotifers, and amoebae. The Endogonales are a unique group in the Zygomycota because some members can form ectomycorrhizal associations with pine roots, while others appear to be saprobic.

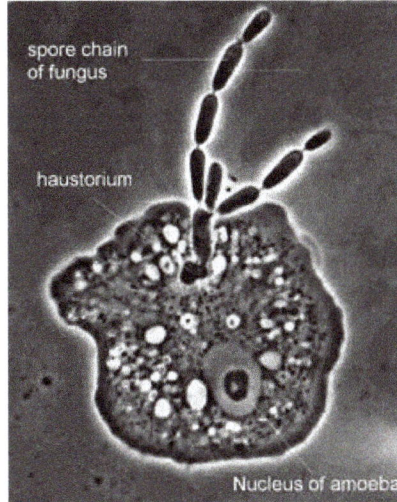

The parasite Amoebophilus simplex (Zoopagales) and its amoeba host. Nutrient transfer occurs through a specialized hypha called a haustorium that enters the amoeba. Spores are produced in chains, and they are engulfed by scavenging amoebae to begin the infection process.

Discussion of Phylogenetic Relationships

The Zygomycota are thought to have diverged from the remaining fungi before the colonization of land by plants 600-1,400 million years ago (Berbee and Taylor 2001; Heckman et al. 2001). Molecular phylogenetic studies place the Zygomycota near the base of the kingdom Fungi, diverging after the Chytridiomycota, the most basal fungal lineage (Bruns et al. 1992; Berbee and Taylor 1993). However, as presently circumscribed, it is uncertain whether the Zygomycota represent a monophyletic group. Studies using SSU rDNA sequence data have generated molecular phylogenies suggesting the Zygomycota may be either para- or polyphyletic (Bruns et al. 1992; Tanabe et al. 2000, 2004). With the recent removal of the Glomales from the Zygomycota (Sch??ler et al. 2001), this phylum is restricted to species which form zygospores through mycelial conjugation, at least in those species where sexual reproduction is known.

Prior to the use of molecular phylogenetics, the Zygomycota were classified into two classes, the Zygomycetes and Trichomycetes. Analyses of SSU rDNA sequences, however, have shown that the Trichomycetes are polyphyletic, comprising what we now know are Ichthyosporean protozoans related to animals and also some true Fungi, the Harpellales, which are nested within the Zygomycetes. Although relationships among the orders are poorly understood, analyses of RPB1 DNA sequences resolved a clade comprising the Kickxellales-Harpellales-Dimargaritales. A morphological synapomorphy for this clade is the possession of a uniperforate septum with a lenticular cavity. A large-scale phylogeny of the Mucorales, using three genes and at least one member of each recognized genus, suggests that several of the largest families and the two largest genera (Mucor and Absidia) are polyphyletic (O'Donnell et al. 2001).

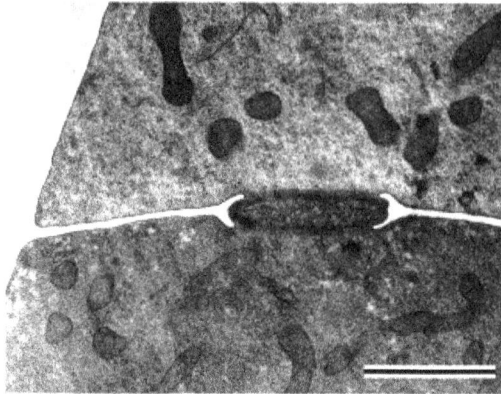

Transmission electron micrograph of vegetative hypha of Kickxella alabastrina (Kickxellales). White line separating the upper from the lower cell is a section of the cross wall or septum. Note the lens shaped plug that lies within the septal pore (lenticular cavity). Scale bar = 0.5 ?m

The Entomophthorales appears to be one of the most distinctive and problematical lineages of Zygomycota for two reasons: 1) SSU rDNA analyses suggest that it may be more closely related to the Blastocladiales (Chytridiomycota) (James et al. 2000; Tanabe et al. 2004), rather than other Zygomycota, and 2) they are morphologically distinct from other Zygomycota in the way their sporangia are formed and in the frequent production of secondary sporangiospores (Cole and Samson 1979; Benny et al. 2001). Phylogenetic placement of one of the most problematic species, Basidiobolus ranarum, is uncertain (Jensen et al. 1998), but a recent phylogenetic analysis using RPB1 sequence data suggests that it is nested within the Zygomycota (Tanabe et al. 2004). However, this species appears to be distinct from the Entomophthorales with which it has been classified traditionally. Although B. ranarum possesses many of the features of other entomophthoralean species, such as forcibly discharged spores, morphologically similar zygospores, and symbiotic associations with insects (Krejzova 1978; Blackwell and Malloch 1989), this species does not appear to group with other Entomophthorales in molecular phylogenetic studies using SSU rDNA sequences (Nagahama et al. 1995; James et al. 2000). Basidiobolus spp. possess centriole-like nuclear-associated organelles (McKerracher and Heath 1985; Cavalier-Smith 1998), however, only members of the Chytridiomycota, the only flagellated true Fungi, possess functional centrioles.

Though controversial, congruent evidence from alpha- and beta-tubulin gene phylogenies support a zygomycete origin of the microsporidia, a group of highly reduced obligate intracellular parasites of a wide variety of animals including humans (Keeling et al. 2000; Keeling 2003). Because several microsporidian species have emerged as major pathogens of immuno-compromised patients over the past two decades, this enigmatic group has received considerable attention recently by the scientific community. Placement of the microsporidia, however, remains controversial.

Spores

You've probably come into contact with spores before. Have you ever walked into your bathroom and noticed a musty smell? And then looked to see a glaze of green and black on the shower curtain? You had mold! You might remember hearing news stories on the dangers of mold and how it can infect your lungs and cause allergies and other nasty illnesses. But it's not just the mold that can make you sick; it's the spores.

Spores are the single-celled reproductive unit of nonflowering plants, bacteria, fungi, and algae. Basically, spores are the babies, except they didn't need a mom and a dad. Not all life forms reproduce sexually. Many, such as fungi and bacteria, reproduce without mating at all. Instead, they produce hardy structures known as spores that are often adapted for dispersal from the main plant or fungus. Spores can last a very long time in some nasty conditions.

What Good are Spores?

Spores are the reproductive structure of the 'lower plants,' plants that don't flower. Fungi, algae, and even some bacteria all form spores when they want to pass their genes on. Think of them like seeds; they are made to grow a new plant and all they need is the proper environment to thrive.

Spores are an asexual form of reproduction; the plant or fungus doesn't need to mate with another plant or fungus to form these particles. A spore is typically a single cell surrounded by a thick cell wall for protection. Once the spores are formed, the organism releases them into the environment to grow and thrive. Spores are often formed through a process called sporogenesis, which just means the production of spores, and is accomplished through mitosis, or cellular reproduction.

Once a spore is produced, it needs to get out into the world where it can grow and thrive. It does this through dispersal adaptations in the spore, which are different features which allow the spore to travel. Some spores are so light they get picked up by the wind and blown to a new place. Other spores ride on the currents of rivers and streams. Still others get shot out into the air by the fungus which made them, or placed in a fragile container that bursts open when touched.

The term "spore" is used to describe a structure related to propagation and dispersal. Zygomycete spores can be formed through both sexual and asexual means. Before germination the spore is in a dormant state. During this period, the metabolic rate is very low and it may last from a few hours to many years. There are two types of dormancy. The exogenous dormancy is controlled by environmental factors such as temperature or nutrient availability. The endogenous or constitutive dormancy depends on characteristics of the spore itself; for example, metabolic features. In this type of dormancy, germination may be prevented even if the environmental conditions favor growth.

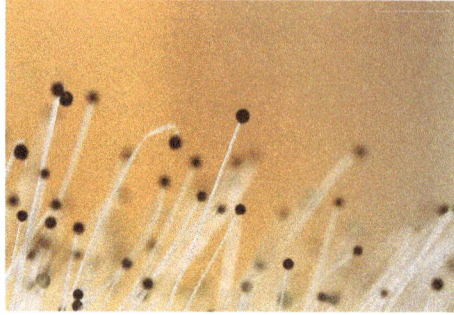

Detail of Sporangia of a Zygomycota species growing on a peach.

Mitospores

In zygomycetes, mitospores (sporangiospores) are formed asexually. They are formed in specialized structures, the mitosporangia (sporangia) that contain few to several thousand of spores, depending on the species. Mitosporangia are carried by specialized hyphae, the mitosporangiophores (sporangiophores). These specialized hyphae usually show negative gravitropism and positive phototropism allowing good spore dispersal. The sporangia wall is thin and is easily destroyed by mechanical stimuli (e.g. falling raindrops, passing animals), leading to the dispersal of the ripe mitospores. The walls of these spores contain sporopollenin in some species. Sporopollenin is formed out of β-carotene and is very resistant to biological and chemical degradation. Zygomycete spores may also be classified in respect to their persistence:

Chlamydospores

Chlamydospores are asexual spores different from sporangiospores. The primary function of chlamydospores is the persistence of the mycelium and they are released when the mycelium degrades. Chlamydospores have no mechanism for dispersal. In zygomycetes the formation of chlamydospores is usually intercalar. However, it may also be terminal. In accordance with their function chlamydospores have a thick cell wall and are pigmented.

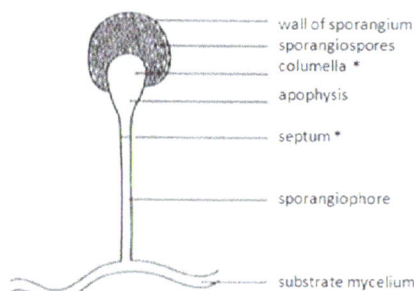

wall of sporangium
sporangiospores
columella *
apophysis
septum *
sporangiophore
substrate mycelium

Sporangium.

Zygophores

Zygophores are chemotropic aerial hyphae that are the sex organs of Zygomycota, except for Phycomyces in which they are not aerial but found in the substratum. They have two different mating types (+) and (-). The opposite mating types grow towards each other due to volatile pheromones given off by the opposite strand, mainly trisporic acid and its precursors. Once two opposite mating types have made initial contact, they give rise to a zygospore through multiple steps.

Zygospore formation is the result of a multiple step process beginning with compatible mating type zygophores growing towards each other. Once contact between the zygophores has been made, their walls adhere to each other, flatten and then the contact site is referred to as the fusion septum. The tips of the zygophore become distended and form what is called the progametangia. A septum develops by gradual inward extension until it separates the terminal gametangia from the progametangial base. At this point the zygophore is then called the suspensor. Vesicles accumulate at the fusion septum at which time it begins to dissolve. A little before the fusion septum completely dissolves, the primary outer wall begins to thicken. This can be seen as dark patches on the primary wall as the fusion septum dissolves. These dark patches on the wall will eventually develop into warty structures that make up the thickness of the zygospore wall. As the zygospore enlarges, so do the warty structures until there are contiguous around the entire cell. At this point, electron microscopy can no longer penetrate the wall. Eventually the warts push through the primary wall and darken which is likely caused by melanin.

Meiosis usually occurs before zygospore germination and there are a few main types of distinguishable nuclear behavior. Type 1 is when the nuclei fuse quickly, within a few days, resulting in mature zygospore having haploid nuclei. Type 2 is when some nuclei do not pair and degenerate instead, meiosis is delayed until germination. Type 3 is when haploid nuclei continue to divide mitotically and then some associate into groups and some do not. This results in diploid and haploid nuclei being found in the germ sporangium.

Microsporidia

The microsporidia are a group of obligate intracellular parasitic fungi. Historically, they have been treated among the protozoa, and as such are often still managed by diagnostic parasitology laboratories. To date, more than 1,200 species belonging to 143 genera have been described as parasites infecting a wide range of vertebrate and invertebrate hosts. Microsporidia, are characterized by the production of resistant spores that vary in size, depending on the species. They possess a unique organelle, the polar tubule or polar filament, which is coiled inside the spore as demonstrated by its ultrastructure.

The microsporidia spores of species associated with human infection measure from 1 to 4 μm and that is a useful diagnostic feature. There are at least 15 microsporidian species that have been identified as human pathogens: Anncaliia (formerly Brachiola) algerae, A. connori, A. vesicularum, Encephalitozoon cuniculi, *E. hellem*, *E. intestinalis*, *Enterocytozoon bieneusi Microsporidium ceylonensis*, *M. africanum*, *Nosema ocularum*, *Pleistophora* sp., *Trachipleistophora hominis*, *T. anthropophthera*, *Vittaforma corneae*, and *Tubulinosema acridophagus*. *Encephalitozoon intestinalis* was previously named *Septata intestinalis*, but it was reclassified as *Encephalitozoon intestinalis* based on its similarity at the morphologic, antigenic, and molecular levels to other species of this genus.

Life Cycle

The infective form of microsporidia is the resistant spore and it can survive for a long time in the environment The number 1. The spore extrudes its polar tubule and infects the host cell The number 2. The spore injects the infective sporoplasm into the eukaryotic host cell through the polar tubule The number 3. Inside the cell, the sporoplasm undergoes extensive multiplication either by merogony (binary fission) or schizogony (multiple fission) The number 4. This development can occur either in direct contact with the host cell cytoplasm (e.g., E. bieneusi) or inside a vacuole termed parasitophorous vacuole (e.g., E. intestinalis). Either free in the cytoplasm or inside a parasitophorous vacuole, microsporidia develop by sporogony to mature spores The number 5. During sporogony, a thick wall is formed around the spore, which provides resistance to adverse environmental conditions. When the spores increase in number and completely fill the host cell cytoplasm, the cell membrane is disrupted and releases the spores to the surroundings The number 6. These free mature spores can infect new cells thus continuing the cycle.

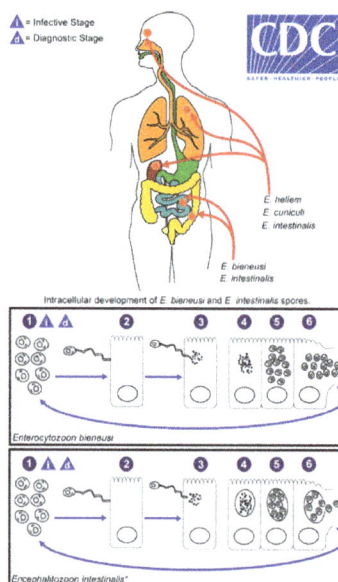

Morphology

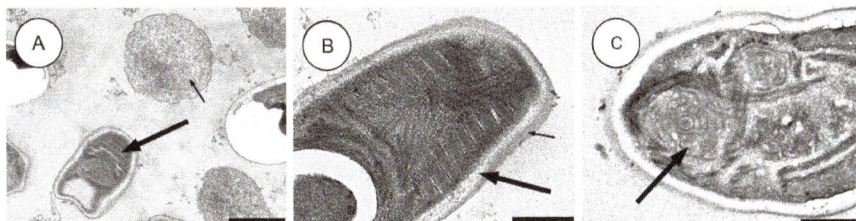

Dictyocoela diporeiae. A, meront and spore; B, spore wall; C, polar filament

Microsporidia lack mitochondria, instead possessing mitosomes. They also lack motile structures, such as flagella.

Microsporidia produce highly resistant spores, capable of surviving outside their host for up to several years. Spore morphology is useful in distinguishing between different species. Spores of most species are oval or pyriform, but rod-shaped or spherical spores are not unusual. A few genera produce spores of unique shape for the genus.

The spore is protected by a wall, consisting of three layers:

- an outer electron-dense *exospore*
- a median, wide and seemingly structureless *endospore*, containing chitin
- a thin internal *plasma membrane*

In most cases there are two closely associated nuclei, forming a *diplokaryon*, but sometimes there is only one. The anterior half of the spore contains a harpoon-like apparatus with a long, thread-like *polar filament*, which is coiled up in the posterior half of the spore. The anterior part of the polar filament is surrounded by a *polaroplast*, a lamella of membranes. Behind the polar filament, there is a posterior *vacuole*.

Infection

In the gut of the host the spore germinates, it builds up osmotic pressure until its rigid wall ruptures at its thinnest point at the apex. The posterior vacuole swells, forcing the polar filament to rapidly eject the infectious content into the cytoplasm of the potential host. Simultaneously the material of the filament is rearranged to form a tube which functions as a hypodermic needle and penetrates the gut epithelium.

Once inside the host cell, a sporoplasm grows, dividing or forming a multinucleate plasmodium, before producing new spores. The life cycle varies considerably. Some have a simple asexual life cycle, while others have a complex life cycle involving multiple hosts and both asexual and sexual reproduction. Different types of spores may be produced at different stages, probably with different functions including autoinfection (transmission within a single host).

Medical Implications

In animals and humans, microsporidia often cause chronic, debilitating diseases rather than lethal infections. Effects on the host include reduced longevity, fertility, weight, and general vigor. Vertical transmission of microsporidia is frequently reported. In the case of insect hosts, vertical transmission often occurs as transovarial transmission, where the microsporidian parasites pass from the ovaries of the female host into eggs and eventually multiply in the infected larvae. *Amblyospora salinaria* n. sp. which infects the mosquito *Culex salinarius* Coquillett, and *Amblyospora californica* which infects the mosquito *Culex tarsalis* Coquillett, provide typical examples of transovarial transmission of microsporidia.

Microsporidia, specifically the mosquito-infecting *Vavraia culicis*, are being explored as a possible 'evolution-proof' malaria-control method. Microsporidian infection of *Anopheles gambiae* (the principal vector of *Plasmodium falciparum* malaria) reduces malarial infection within the mosquito, and shortens the mosquito lifespan. As the majority of malaria-infected mosquitoes naturally die before the malaria parasite is mature enough to transmit, any increase in mosquito mortality through microsporidian-infection may reduce malaria transmission to humans.

Clinical

Microsporidian infections of humans sometimes cause a disease called microsporidiosis. At least 14 microsporidian species, spread across eight genera, have been recognized as human pathogens. These include *Trachipleistophora hominis*.

As Hyperparasites

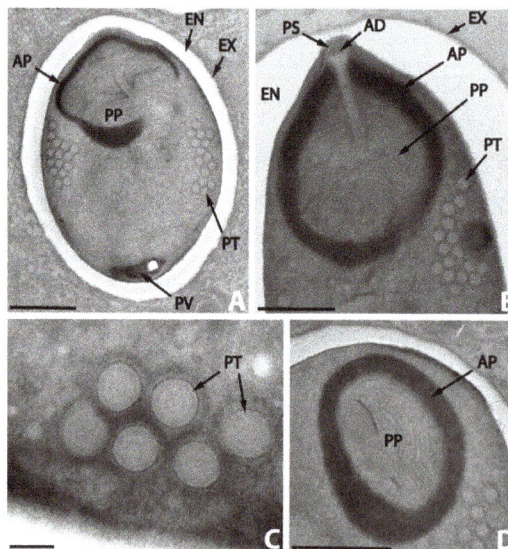

A hyperparasitic microsporidian, *Nosema podocotyloidis*, a parasite of a digenean which is itself a parasite of a fish

Microsporidia can infect a variety of hosts, including hosts which are themselves parasites. In that case, the microsporidian species is an hyperparasite, i.e. a parasite of a parasite. As an example, more than eighteen species are known which parasitize digeneans (parasitic flatworms). These digeneans are themselves parasites in various vertebrates and molluscs. Eight of these species belong to the genus *Nosema*.

Genomes

Microsporidia have the smallest known (nuclear) eukaryotic genomes. The parasitic lifestyle of microsporidia has led to a loss of many mitochondrial and Golgi genes, and even their ribosomal RNAs are reduced in size compared with those of most eukaryotes. As a consequence, the genomes of microsporidia are much smaller than those of other eukaryotes. Currently known microsporidial genomes are 2.5 to 11.6 Mb in size, encoding from 1,848 to 3,266 proteins which is in the same range as many bacteria.

Horizontal gene transfer (HGT) seems to have occurred many times in microsporidia. For instance, the genomes of *Encephalitozoon romaleae* and *Trachipleistophora hominis* contain genes that derive from animals and bacteria, and some even from fungi.

Classification and Taxonomy

The scientific classification of Microsporidia has evolved through time with growing scientific research in the area, and the specifics are still currently debated. Initially thought to be a protozoan (kingdom Protista), recent studies using DNA techniques indicate phylum Microspora should be classified under the Fungi kingdom or at least as a sister kingdom to Fungi. The class, order and family within the Microspora phylum are also frequently revised and debated (traditional taxomony given at the family level bloew). Traditionally, species were identified by observing the physical characteristics of the spore, life cycle and relationship with the host cell. However, recent scientific studies using genetic tools (namely ribosomal RNA sequencing) have challenged this approach and suggest genetic markers a more correct method for scientific classification. More research is still needed to better understand the origins of microspora and of individual species. Despite all this, there are now over 1200 species identified in 143 genera. Currently, at least 14 species in 8 genera are known to infect humans.

Family	Genera	Species
Nosematidea	*Brachiola*	*B. algerae, B. vesicularum*
Encephalitozoonidea	*Encephalitozoon*	*E. cuniculi, E. hellem, E. intestinalis (syn. Septata intestinalis).*
Enterocytozoonidea	*Enterocytozoon*	*Enterocytozoon bieneusi,*

Microsporidea	*Microsporidium*	*M. ceylonensis, M. africanum*
Nosematidea	*Nosema*	*N. ocularum, N. connori (syn. B connori)*
Pleistophoridea	*Pleistophora*	*Sp.*
Pleistophoridea	*Trachipleistophora*	*T. hominis, T. anthropophthera,*
Nosematidea	*Vittaforma*	*Vittaforma corneae (syn. Nosema corneum)*

References

- Hawksworth DL. (2013). Ascomycete Systematics: Problems and Perspectives in the Nineties. Springer. p. 116. ISBN 978-1-4757-9290-4

- Lin, X.; Hull, C. M.; Heitman, J. (April 2005). "Sexual reproduction between partners of the same mating type in Cryptococcus neoformans". Nature. 434 (7036): 1017–21. Bibcode:2005Natur.434.1017L. doi:10.1038/nature03448. PMID 15846346

- "Part 1- Virae, Prokarya, Protists, Fungi". Collection of genus-group names in a systematic arrangement. Retrieved 30 June 2016

- Silar P (2016). "Protistes Eucaryotes: Origine, Evolution et Biologie des Microbes Eucaryotes". HAL: 462. ISBN 978-2-9555841-0-1

- Powell; Letcher (2015). A new genus and family for the misclassified chytrid, Rhizophlyctis harderi (in press). Mycologia. doi:10.3852/14-223. Retrieved 2016-08-23

- Michod, R. E.; Bernstein, H.; Nedelcu, A. M. (May 2008). "Adaptive value of sex in microbial pathogens". Infection, Genetics and Evolution. 8 (3): 267–85. doi:10.1016/j.meegid.2008.01.002. PMID 18295550

- Corliss JO, Levine ND (1963). "Establishment of the Microsporidea as a new class in the protozoan subphylum Cnidospora". The Journal of Protozoology. 10 (Suppl.): 26–27

- Freeman, K.R. "Evidence that chytrids dominate fungal communities in high-elevation soils". pnas.org. Retrieved 28 October 2013

3

Anatomy and Morphology of Fungus

Fungi generally grow as hyphae. These are thread-like cylindrical structures. This chapter contains detailed description of various fungal structures, such as hypha, sporocarp, conidium, crozier and veil. The topics elaborated in this chapter will help in gaining a better perspective about the anatomy and morphology of fungi.

Hypha

Hyphae are comprised of hypha, which are the long filamentous branches found in fungi and actinobacteria. Hyphae are important structures required for growth in these species, and together, are referred to as mycelium.

Hyphae Structure

Each hypha is comprised of at least one cell encapsulated by a protective cell wall typically made of chitin, and contain internal septa, which serve to divide the cells. Septa are important as they allow cellular organelles (e.g., ribosomes) to pass between cells via large pores. However, not all species of fungi contain septa. The average hyphae are approximately 4 to 6 microns in size.

Hyphae Growth

Hyphae growth occurs by extending the cell walls and internal components from the tips. During tip growth, a specialized organelle called the spitzenkörper, assists in the formation of new cell wall and membrane structures by harboring vesicles derived from the golgi apparatus and releasing them along the apex of the hypha. As the spitzenkörper moves, the tip of the hypha is extended via the release of the vesicle contents, which form the cell wall, and the vesicle membranes, which create a new cell membrane. As the hypha extends, new septa can be created to internally divide the cells. The characteristic branching of hyphae is the result of the formation of a new tip from a hypha, or the division of a growing tip.

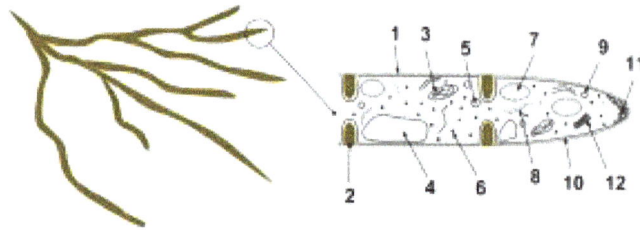

(1- Hyphal wall 2- Septum 3- Mitochondrion 4- Vacuole 5- Ergosterol crystal 6- Ribosome 7- Nucleus 8- Endoplasmic reticulum 9- Lipid body 10- Plasma membrane 11- Spitzenkörper/growth tip and vesicles 12- Golgi apparatus)

Hyphae Function

Hyphae are associated with multiple different functions, depending on the specific requirements of each fungal species. The following are a list of the most commonly known hyphae functions:

Nutrient Absorption from a Host

Some hyphae of parasitic fungi are specialized for nutrient absorption within a specific host. These hyphae have specialized tips called haustoria, which penetrate the cell walls of plants or tissues of other organisms in order to obtain nutrients.

Nutrient Absorption from Soil

Some fungal species (e.g., *mycorrihizae*) have developed a symbiotic relationship with vascular plant species. The fungi forms specialized hyphae called arbuscules, which can be found in the roots or phylum of vascular plants, and function to absorb nutrients and water from the soil. In this manner, the hyphae aid the plants by increasing its access to nutrients in the soil while facilitating its own growth.

Trapping Structures

In some fungal species, hyphae have evolved into specialized nematode-trapping structures, using nets and ring structures to trap nematode species.

Nutrient Transportation

Several fungal species exhibit hyphae composed of chord-like structures, termed mycelial chords, which are used by fungi (e.g., lichens and mushrooms) to transport nutrients across great distances.

Hyphae Classification

In general, hyphae can be classified based on the following traits:

Hyphae Characteristics

Hyphae characteristics are an important method of classifying various fungal species. There are three main hyphae characteristics:

- Binding: Binding hyphae have a thick cell wall and are highly branched.

- Generative: Generative hyphae have a thin cell wall, a large number of septa, and are typically less differentiated. Generative hyphae may also be contained within other materials (e.g., gelatin or mucilage) and can also develop structures used in reproduction. All fungal species typically contain generative hyphae.

- Skeletal: Skeletal hyphae contain a long and thick cell wall with few septa. Skeletal hyphae can also be of a fusiform subtype, with a swollen midsection surrounded by tapered ends.

Hyphae Composition

Fungal species are also further classified based on the hyphal systems they contain. There are four general subtypes:

- Monomitic: While virtually all fungal species contain generative hyphae, those with only exhibit this type are referred to as monomitic (e.g., agaric mushrooms).

- Dimitic: A species that contains generative hyphae in addition to one other type of hyphae. The most common combination of dimitic fungi is generative and skeletal.

- Trimitic: Species which contain all three types of hyphae (generative, binding, and skeletal).

- Sarcodimitic and sarcotrimitic: Sarcodimitic hyphae are fusiform skeletal hyphae bound to generative hyphae. Sarcotrimitic species contain fusiform skeletal hyphae, as well as binding and generative hyphae.

Hyphae Refraction

Under a microscope, the appearance of oily or granular hyphae under a microscope is termed gloeoplerous. This term is also used to further classify the hyphae of various species.

Cell Division

Hyphae can be classified based on the presence of internal septa (septate versus aseptate species). Hyphae can also be distinguished from species which produce pseudohyphae via cell division. Pseudohyphae is a form of incomplete cell division, in which the dividing cells do not separate. There are several yeast species which produce such pseudohyphae.

Sporocarp

The descriptions of the mushroom and the underground mycelium encapsulate the essence of all the (macro) fungi. Basically, there's an out-of-sight mycelium and visible spore producing structures - called fruiting bodies or sporocarps.

You'll already be familiar with some of them. For example, most people have seen puffballs on the ground and bracket fungi growing out of trees. But remember whenever you see a mushroom growing from wombat dung, a puffball growing on a grassy oval or a bracket fungus growing from a gum tree - you're seeing just the sporocarp. Its sole function is to produce and disperse spores.

Out of sight there's a mycelium in the wombat dung, under the grassy oval or in the trunk of the gum tree - secreting enzymes to break down organic matter in the dung, soil or tree trunk.

Strictly speaking, it is incorrect to call a mushroom a fungus, since we are only looking at part of the fungus, but people commonly refer to mushrooms, puffballs, etc. as 'fungi' - which is understandable since these are the only parts we usually see.

These are all terms for the part of the fungus that produces its spores. Carpophore, sporocarp, sporophore, and fruiting body are the most general terms, referring to any such organ.

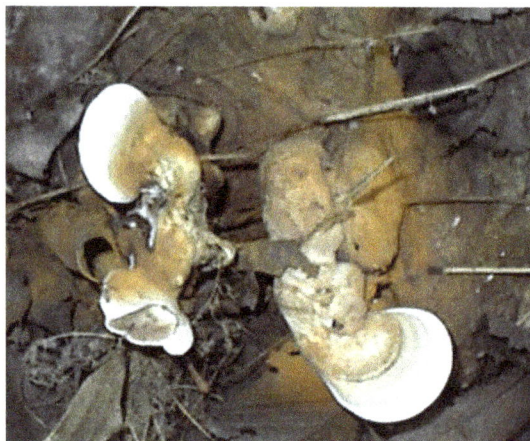

Photo of Ganoderma applanatum by John Denk

A macroscopic conidia-bearing fruiting body is currently called a conidioma, but yes, it also used to be called a conidiocarp. The microscopic conidia-bearing structure, a condiophore, is more analogous to the microscopic cells that bear sexual fungal spores, basidia and asci, or perhaps to the hymenium. In the picture, you can see that these fruiting bodies of Ganoderma applanatum have produced so many of their reddish brown spores that the leaves around them are thickly dusted with them.

Pileus Mycology

The pileus is the cap of the mushroom. A mushroom with a pileus is said to be pileate. A mushroom lacking a pileus, that is consisting of just a fertile surface with its back attached to or intergrown with the substrate, is said to be resupinate. Resupinate fungi usually grow on a substrate that is horizontal, like a fallen log.

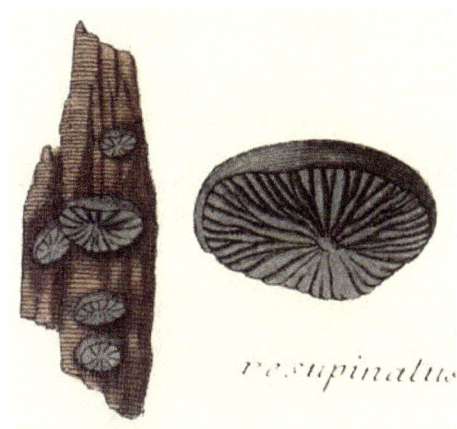

Image of Resupinatus applicatus from Christian Gottfried Daniel Nees von Esenbeck (1816 - 1817) *Das System der Pilze und Schwämme*

Effused-reflexed is a term used to describe a mushroom on a vertical surface that is partially resupinate and partially pileate, as in the picture. The pileate portion of an effused-reflexed fruiting body is necessarily sessile. The part that's sticking out is sometimes called the reflexed portion, and the resupinate part is sometimes called the effused portion.

Photo of Polyporaceae by John Denk

The edge of the cap is called the margin, and the center of the top surface of the cap is called the disk (or disc). This specimen of Hygrophorus russula has a white margin, and the cuticle darkens considerably at the disk.

Hygrophorus russula from Christian Gottfried Daniel Nees von Esenbeck (1816 - 1817) Das System der Pilze und Schwämme

In some species, the margin is inrolled, and this can be a helpful feature in identifying the mushroom. The term incurved is sometimes used in this sense.

Clitocybe nebularis from A. M. Hussey (1847 - 1855) Illustrations of British mycology

If the margin extends beyond the gills or tubes, the mushroom is said to have a sterile margin.

Photo of Agaricus bitorquis by John Denk

If the margin is wavy, it is called sinuate, just like a gill margin or attachment.

Photo of Clitocybe nuda by Leon Shernoff

If the margin is looks like it has irregular, large, blunt, blobby extensions, it is said to be lobed. The margin of a mushroom often becomes lobed in age, as in this picture.

Image of Clitocybe dealbata from A. M. Hussey (1847 - 1855) Illustrations of British mycology

If the margin is broken up into many smaller, sharper protrusions that look like the cap has been torn slightly in order to make them, the margin is called lacerate. As with the lobed condition, a margin often becomes lacerate as the mushroom ages; but here it starts out quite young.

Image of Leptonia porphyrophaea from A. M. Hussey (1847 - 1855) Illustrations of British mycology

Classification

Pilei can be formed in various shapes, and the shapes can change over the course of the developmental cycle of a fungus. The most familiar pileus shape is hemispherical or *convex*. Convex pilei often continue to expand as they mature until they become flat.

Many well-known species have a convex pileus, including the button mushroom, various *Amanita* species and boletes.

Some, such as the parasol mushroom, have distinct bosses or umbos and are described as *umbonate*. An umbo is a knobby protrusion at the center of the cap. Some fungi, such as chanterelles have a funnel- or trumpet-shaped appearance. In these cases the pileus is termed *infundibuliform*.

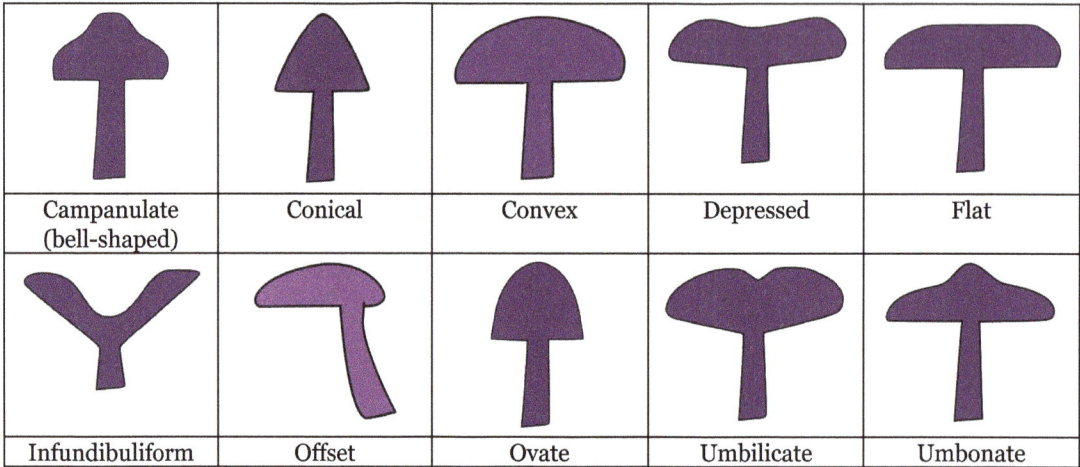

Campanulate (bell-shaped)	Conical	Convex	Depressed	Flat
Infundibuliform	Offset	Ovate	Umbilicate	Umbonate

Basidiocarp

In fungi, a basidiocarp, basidiome or basidioma (plural: basidiomata) is the sporocarp of a basidiomycete, the multicellular structure on which the spore-producing hymenium is borne. Basidiocarps are characteristic of the hymenomycetes; rusts and smuts do not produce such structures. As with other sporocarps, epigeous (above-ground) basidiocarps that are visible to the naked eye (especially those with a more or less agaricoid morphology) are commonly referred to as mushrooms, while hypogeous (underground) basidiocarps are usually called false truffles.

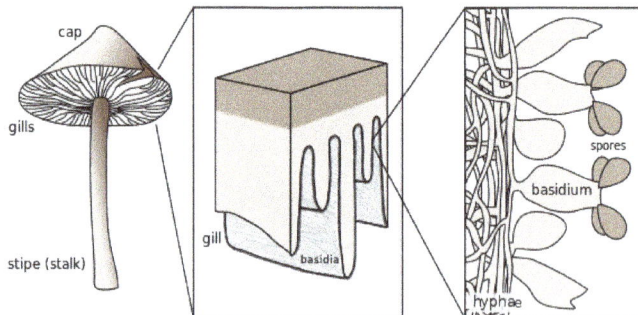

Schematic of a typical basidiocarp, showing fruiting body, hymenium and basidia

Structure

All basidiocarps serve as the structure on which the hymenium is produced. Basidia are found on the surface of the hymenium, and the basidia ultimately produce spores. In its simplest form, a basidiocarp consists of an undifferentiated fruiting structure with a hymenium on the surface; such a structure is characteristic of many simple jelly and club fungi. In more complex basidiocarps, there is differentiation into a stipe, a pileus, and/or various types of hymenophores.

Types

Basidiocarps are classified into various types of growth forms based on the degree of differentiation into a stipe, pileus, and hymenophore, as well as the type of hymenophore, if present.

Basidiocarps of *Amanita muscaria*, an agaric.

Basidiocarps of *Ramaria rugosa*, a coral fungus

Basidiocarps of *Craterellus tubaeformis*, a cantharelloid fungus.

Growth forms include:

- jelly fungus – fruiting body is jelly-like.

- club fungus and coral fungus – erect fruiting body without a distinct stalk and cap, either unbranched (club fungus) or profusely branched (coral fungus).

- polypore – underside of the fruiting body usually consists of tubes; otherwise very variable, usually wood-inhabiting

- tooth fungus or hydnoid fungus - underside of the fruiting body composed of spines or teeth

- corticioid fungus - the underside of the fruiting body is usually smooth or with spines (vs. hydnoid fungi) but not poroid nor gilled; typically effused without caps

- cantharelloid fungus – fruiting body with shallow fold-like gills running over most of the lower surface of the fruiting body and not much differentiation between the stalk and cap.

- gasteromycete or "gastroid fungus" – fruiting body has a ball-like shape and in which the hymenophore has become entirely enclosed on the inside of the fruiting body.

- false truffle – like a gasteromycete, however, but with a hypogeous (underground) fruiting body.

- secotioid fungus – like a gasteromycete, but with a stalk. Thought to be an evolutionarily intermediate stage between a gasteromycete and an agaric.

- agaric or gill fungi – fruiting body with caps, gills, and (usually) a stalk.

- bolete – fleshy fruiting body with a cap, a stalk, and tubes on the underside.

Basic divisions of Agaricomycotina were formerly based entirely upon the growth form of the mushroom. Molecular phylogenetic investigation (as well as supporting evidence from micromorphology and chemotaxonomy) has since demonstrated that similar types of basidiomycete growth form are often examples of convergent evolution and do not always reflect a close relationship between different groups of fungi. For example, agarics have arisen independently in the Agaricales, the Boletales, the Russulales, and other groups, while secotioid fungi and false truffles have arisen independently many times just within the Agaricales.

Ascocarp

Ascocarp, also called ascoma, plural ascomata, fruiting structure of fungi of the phylum Ascomycota (kingdom Fungi). It arises from vegetative filaments (hyphae) after sexual reproduction has been initiated. The ascocarp (in forms called apothecium, cleistothecium [cleistocarp], or perithecium) contain saclike structures (asci) that usually bear four to eight ascospores. Apothecia are stalked and either disklike, saucer-shaped, or cup-shaped with exposed asci. The largest known apothecium, produced by *Geopyxis cacabus,* has a stalk 1 metre (40 inches) high and a cup 50 centimetres (20 inches) across. Cleistothecia are spherical and must rupture or disintegrate to release their ascospores. Perithecia are globular or flask-shaped with an apical opening for discharge of ascospores.

'Following is the description of the range and development of ascocarps: '

The fructifications or fruit bodies or ascocarps of the Ascomycetes are structures containing asci surrounded by, or enclosed within, protective tissue—peridium.

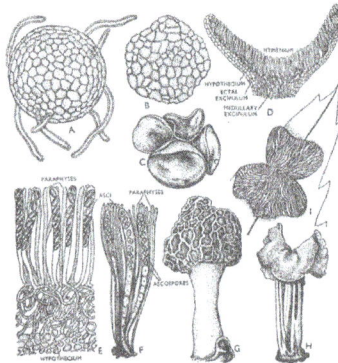

Fig. 205. Various types of ascocarps. A. *Erysiphe* sp. Cleistothecium with appendages. B. *Eurotium* sp. Cleistothecium without appendages. C. *Peziza* sp. Apothecia. D. Diagram of a section through an apothecium showing structural details. E. Portion of a section through an apothecium showing hypothecium, asci and paraphyses. F. Portion of hymenium with asci, ascospores and paraphyses. G. *Morchella* sp. Stipitate apothecium with sponge-like pileus. H. *Helvella* sp. Stipitate apothecium with saddle-shaped pileus. I. *Rhytisma* sp. Tar-like stroma with radiate wrinkles under which apothecia are formed

Their general structure, conditions of growth and shape are extremely variable. Asco-carps may occur singly or in groups, and may be superficial, erumpent (bursting through the substratum) or embedded in the substratum or host tissues. The origin and distribution of hymenium are also different in them.

The hymenium may consist of asci only or the asci may be interspersed with sterile hy-phal elements, paraphyses (sing, paraphysis) arising in the ascogenous layer.

The paraphyses may be simple or branched, sometimes coloured, sometimes inflated at the apex (capitate), or at intervals along their length (moniliform), and with or without septation. Paraphyses are said to space out the asci and to secrete mucilage which is concerned in spore discharge mechanisms. The fructifications may be flat expansions, crusts, rounded, cup-like, flask-like, alveolar, or erect club-like, etc.

But the most common and widely encountered forms are:

(i)   A more or less rounded structure, completely closed, and having hymenium enclosed within the wall, a cleistothe- dum (pl. cleistothecia);

(ii) A saucer- or cup-shaped ascocarp with wide open hymenium, an apothecium (pl. apothecia);

(iii) A flask-like ascocarp designated as a perithecium (p1. perithecia) perforated at the apex by a pore known as ostiole. Sometimes the ostiole is prolonged into a beak —rostrum and usually it is lined with hair-like hyphal structures, periphyses (sing, periphysis) directed towards the opening and assisting in expelling asci or ascospores.

A perithecium has its own distinct peridium of specialized tissue and may or may not be embedded in a stroma and is thus designated as stromatic or non-stromatic perith-ecium. In addition to these forms of ascocarp, there are others which are not so widely encountered.

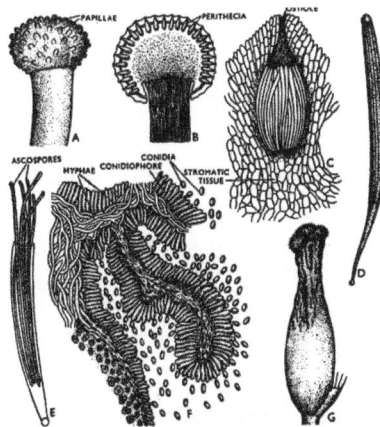

Fig. 246. *Claviceps purpurea*. A. Papillae on the surface of the sphaeridium. B. Longitudinal section of a sphaeridium showing single layer of perithecia. C. A perithecium. D. An ascus. E. Ascus dis-charging ascospores. F. Conidial, 'honey-dew' stage on the ovary. G. Immature sclerotium.

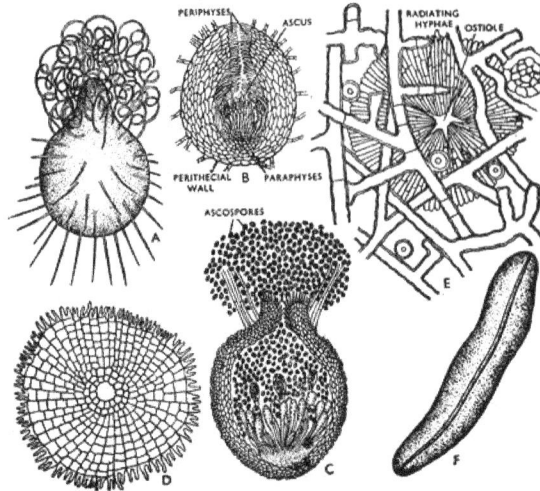

Fig. 206. Various types of ascocarps. A–C. *Chaetomium* sp. A. Perithecium with numerous long hairs on its surface. B. Section through the perithecium showing structural details. C. Section through the perithecium showing asci and dispersal of ascospores. D. *Microthyrium microscopicum*. Shield-shaped, dimidiate ascostroma with an opening. E. *Asterina* sp. Thyriothecium composed of radiating hyphae. F. *Glonium* sp. Boat-shaped ascocarp (hysterothecium) with a long slit-like opening parallel to the long axis and extending almost through the entire length of the fruit body.

They may be black, carbonous or leathery, boat-shaped or branched, etc. Some of them are designated as:

(i)   A hysterothecium (pl. hysterothecia), an elongated ascocarp closed when young, but opens at maturity by a long slit following a line of dehiscence parallel to the long axis extending almost the entire length of the ascocarp; and

(ii) A thyrkn thecium (pl. thyriothecia), also known as catathecium or catothecium, is an inverted saucer-shaped ascocarp having wall more or less radial in structure. The hysterothecium is often considered to be an intermediate form between an apothecium and a perithecium.

Again there are ascocarps that are developed as locules or cavities, especially ascigerous cavities, in a stroma where asci are directly borne. These locules are also designated as pseudothecia (sing, pseudothecium) or pseudoperithecia or as ascostromata (sing, ascostroma). A pseudothecium may be uni- or multilocular in which asci are borne in the base of the locule. The asci may or may not be mixed with hyphal structures which are attached both to the roof and to the base of the locule.

These hyphal structures having no free ends are known as pseudo- paraphyses (sing, pseudoparaphysis), also designated as paraphysoids. The cavity of a pseudothecium may or may not be developed before the formation of asci and paraphysoids. In a pseudothecium, there is no special wall similar to what is present in a perithecium.

Only a cavity is present within the stroma in which asci are located. The development of a regular ostiole is absent in a pseudothecium, instead a pore or slit is formed at a point above the asci where the stromatic tissue is dissolved. Through this pore the ascospores escape.

Besides the above forms, the fructifications may be hypogean and closed, as in the genus Tuber. Ascospores are liberated by the decay or by the breaking of the fructifications by animals.

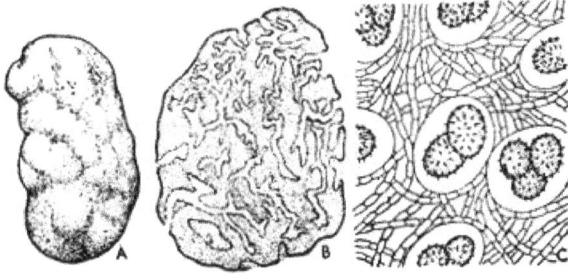

Fig. 207. Nature of ascocarp of *Tuber* sp. A. An entire ascocarp. B. Section through an ascocarp. C. Asci containing ascospores.

The development of fructifications in the Ascomycetes, up to the present time, has been closely studied in a comparatively few species. In general, die process is initiated by the stimulus produced during sexual reproduction.

In some Ascomycetes the ascocarp tissues are differentiated around the sex organs after plasmogamy, while in others the ascocarp tissues form first, forming a stroma within which the sex organs appear later.

In a fructification the hymenium and sterile tissue originate from different hyphal elements which in some cases, may be very well differentiated from the very inception or at best in a very early stage of the fructification. Whereas, in others they are not clearly distinguishable.

Usually the development of hymenium and the steriletissue is simultaneous. Immediately after plasmogamy, the ascogonium gives rise to structures known as ascogenous hyphae from which asci are developed. The ascogenous hyphae are generally broader than the associated vegetative hyphae and have thinner walls, and dense easily stained contents.

They grow out radially to form a plate or hollow disk in the lower part of the fructification, but they may also ramify throughout the central tissue of the ascocarp. Along with this process, profuse growth of sterile hyphae takes place from the adjacent hyphal cells.

In a mature fructification the elements producing asci and the sterile hyphae become so intimately united and interwoven with one another that it is often difficult to separate or distinguish them.

But most fructifications differentiate into two, and in more highly developed ones three, Hyphal systems:

(i)   Tissues derived from the ascogonium comprising of ascogenous hyphae and asci;

(ii)   Vegetative protective tissue developed from the surrounding mycelium; and

(iii) Secondary protective tissue formed under stimulus from the ascogonium. All the tissues and structures enclosed by the peridium or other boundary layer of the ascocarp are known collectively as the centrum. Thus the centrum includes sex organs, ascogenous hyphae, asci and sterile tissues. The structural details and development of three widely encountered forms of fructification are discussed separately.

Cleistothecium:

A cleistothecium is a rounded, completely closed ascocarp which has no natural opening but brusts irregularly at maturity, or along sutures in the peridium. The cleistothecial wall may or may not be provided with outgrowths on its external surface. These outgrowths are known as appendages. The appendages when present serve character of taxonomic interest.

In a cleistothecium the hymenium may be distributed at different levels or at a particular area within the ascocarp. The number of asci in a cleistothecium may be one or more.

Again the asci may be persistent or evanescent. The number of asci in a cleistothecium and whether the asci are persistent or evanescent are also characters on the basis of which genera of certain fungi are separated. The mature ascospores are released from a cleistothecium by external influences which cause the rupture or decay of its wall.

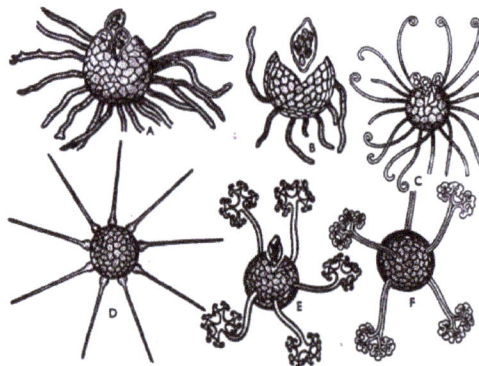

Fig. 243. Taxonomic characteristics of the different genera of the Erysiphaceae. A. *Erysiphe*. Cleisto-thecium with several asci and mycelioid appendages. B. *Sphaerotheca*. Cleistothecium with one ascus and mycelioid appendages. C. *Uncinula*. Cleistothecium with several asci and appendage tips coiled. D. *Phyllactinia*. Cleistothecium with several asci and appendages with a bulbous base. E. *Podosphaera*. Cleistothecium with one ascus and appendages with tips dichotomously branched. F. *Microsphaera*. Cleistothecium with several asci and appendage tips dichotomously branched.

In Podosphaera, the process of cleistothecium development is very simple. Two hyphal branches put out short protuberances at the same time and are delimited by antheridial branch. The antheridial branch remains closely applied to the ascogonium and its upper extremity bends over and covers the apex of the ascogonium, and is delimited by a transverse wall, forming a short nearly isodiametric cell, the antheridium.

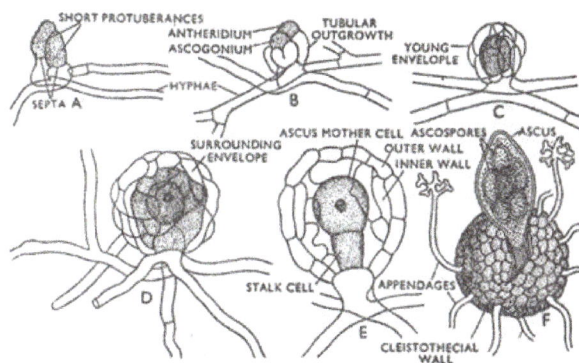

Fig. 208. Development of cleistothecium with ascus and ascospores in *Podosphaera* sp. A. Short protuberances cut off by septa. B. Antheridium and ascogonium surrounded by tubular outgrowths. C–D. Stages in development of surrounding envelope. E. Differentiation of ascus mother cell and stalk cell. F. Mature cleistothecium with appendages and ascus bearing ascospores.

After plasmogamy the ascogonium divides by transverse wall into two cells, an upper which becomes the solitary ascus and subsequently produces eight spores, and a lower which is a stalk cell bears the ascus.

The wall of the cleistothecium also begins to develop at the same time with the ascus. Tubular outgrowths appear close round the base of the ascogonium on the hyphae which bear it and the antheridial branch.

These tubular outgrowths grow up round the ascogonium and in close contact with it and in close lateral contact with one another and with the antheridial branch, till they all meet together above the apex of the ascogonium.

Each one of the tubular outgrowths then divides by one or two transverse walls, so that the ascogonium is surrounded by an envelope formed of a single layer of many cells.

The cells then increase in size in the surface-direction, their walls thicken and assume a dark-brown colour and they thus form the outer wall of the cleistothecium from the outer surface of which number of hyphae with delicate ramifications are developed which ultimately mature into appendages.

Branches from the inner surface of the cells of the outer wall ramify and develop into a dense parenchyma-like weft resulting in the formation of the inner wall of the cleistothecium.

Apothecium:

An Apothecium is a cup- or saucer-shaped ascocarp in which the hymenium remains wide open. The hymenium is uniformly continuous and is developed lining the wide open surface of the apothecium. It is made up of asci and paraphyses. The paraphyses may be as long as the asci, longer or somewhat shorter. They are extremely variable in shape and may be septate or aseptate.

The tips of the paraphyses are usually unbranched. In cases where the tips are branched, the tips of the branches may unite above the asci forming a layer known as epithecium (pl. epithecia). The rest of the tissue of the apothecium is composed of interwoven hyphae.

The thin layer of interwoven hyphae immediately below the hymenium is the hypothecium whose hyphal elements are rather less dense than the rest of the tissue of the apothecium. The fleshy part of the apothecium which supports the hypothecium and the hymenium is the excipulum.

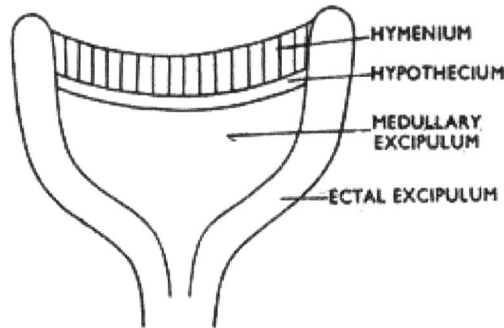

Diagram of an apothecium in section showing arrangements of tissues.

The excipulum consists of two parts:

(i)   The ectal excipulum is the outer layer of the apothecium, and

(ii) The medullary excipulum, forms the inner portion.

Korf (1958) defined in details the types of tissue found in the excipulum. According to him, the ectal excipulum is short-celled, in which the individual hyphae are not recognizable; and the medullary excipulum is with long cells, in which. The component hyphae are visible.

The short-celled tissue he subdivided into three types as follows:

(a)   Cells globose with intercellular spaces—textura globuloaa;

(b)   Cells polyhedral by mutual pressure, no intercellular spaces—textura angularis;

(c)   Cells rectangular in section— long-celled tissue he divided into four subgroups. Long-celled tissue of medullary excipulum may have- (i) hyphae running in all directions;

(a) Hyphal walls not united, usually with distinct interhyphal spaces— textura intricata;

(b) Hyphal walls united, with distinct interhyphal spaces—textura epidermoidea; and (ii) hyphae are more or less parallel:

(c) Hyphae with strongly thickened walls, cohering—textura oblita;

(d) Hyphae without thickened walls, not cohering —textura porrecta.

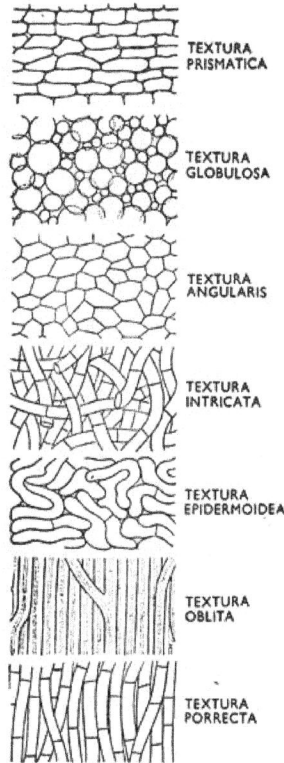

Fig. 210. Types of tissue in the excipulum.

Since the hymenium is wide open, the discharge of ascospores from the asci is rather simple and often clouds of spores are seen on the surface of mature apothecium.

The apothecia are white or brilliantly coloured red, yellow, or orange. Some may be brown. A few are black. The apothecia may be stipitate or non-stipitate, that is, with or without any stalk or stipe respectively.

Besides the typical cups or saucers, there are various other forms of apothecia:

(i) Apothecium with a thick stalk and a cap or pileus on which the pitted or ridged hymenium looks like a sponge;

(ii) Apothecium having heavily ridged short or long stalk and the pileus may be quite large being saddle- shaped or convoluted and brain like;

(iii) Aothecium may be tongue-, club-, or fan-shaped with long stalk;

(iv) oFlat, circular, black, tar-like stroma which bear radiate wrinkles under which apothecia are formed.

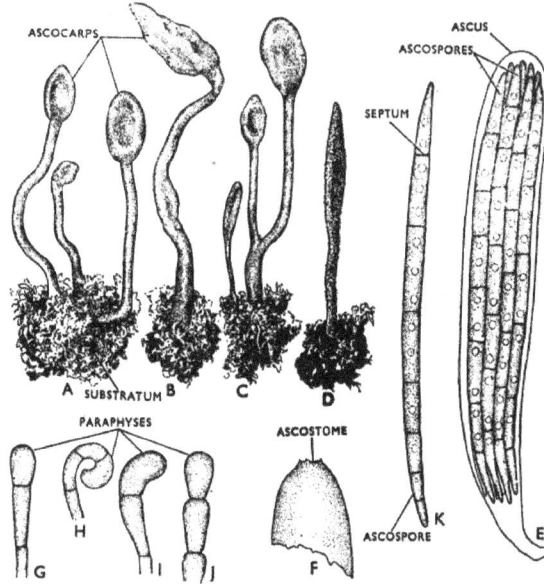

Fig. 239. *Geoglossum* sp. A–D. Ascocarps of varying shape and size. E. An ascus bearing septate ascospores. F. Portion of an ascus showing ascostome. G–J. Different types of paraphyses. K. An ascospore showing nature of septation.

The development of apothecium in Pyronema confluens was imperfectly described by De Bary for the first time in 1863. Tulasne also added some information after De Bary. Later on Kihlmann studied in further details. The details of gametangial contact; plasmogamy; development of dikaryons and ascogenous hyphae; crozier formation; karyogamy; meiosis; and formation of asci and pores.

Simultaneous with all these processes from gametangial contact to the development of ascospores, the vegetative cells at the base of the ascogonium and antheridium put out a growth of vegetative mycelium which surrounds the ascogenous hyphae and ultimately the asci. The two elements form a compact structure to constitute the apothecium.

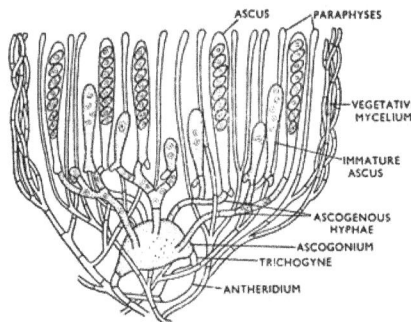

Fig. 211. *Pyronema confluens*. Diagrammatic representation of the development of different parts of the apothecium.

Corner (1929) defined three types of development of apothecia:

(i)   Angiocarpic development, in which the tissues of the excipulum initially grow over and protect the developing hymenial layer;

(ii) Gymnocarpic development, in which the hymenium is exposed throughout its development; and

(iii) Hemiangiocarpic development an intermediate condition in which the hymenium is partially covered. However Corner did not attribute any significance to these types of development.

Perithecium:

The perithecia vary in shape from globose to pear-shaped or elongated to flask-shaped with long or short neck. They may be associated with stroma and are known as stromatic perithecia or free of any stromatic connection, non-stromatic perithecia. Again, the stroma of stromatic perithecia may be only of fungal hyphae or it may be composed of a mixture of fungal hyphae and host tissue.

The stroma may not necessarily surround the perithecia, but may instead be reduced to a shield-like clypeus around the ostiole, or to a subiculum of stromatic interwoven hyphae below the perithecium.

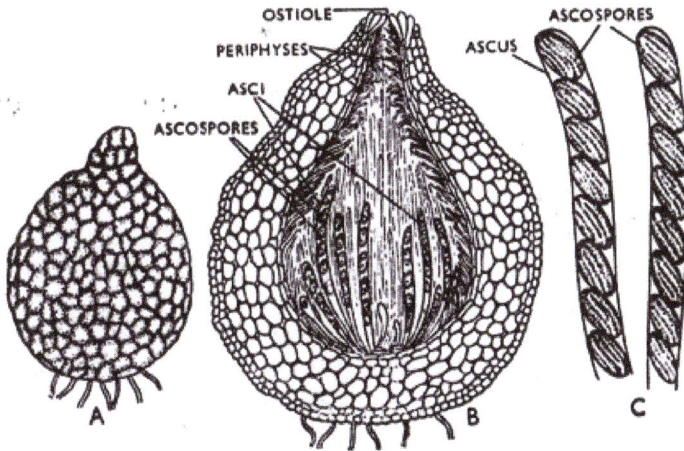

Fig. 249. *Neurospora* sp. A. An entire perithecium. B. Sectional view of a perithecium. C. Asci with ascospores.

A perithecium is furnished in the full-grown state with a schizogenous pore, ostiole through which asci or ascospores are discharged. It is a conspicuous feature of the perithecium. The ostiolar canal is lined by short hair-like growths known as peri- physes (sing, periphysis).

A perithecium, stromatic or non-stromatic has a definite wall of its own with a true ostiole. In a stromatic perithecium, the differentiation of wall may not be sometimes very distinct.

The perithecial wall is usually formed of a dense hyphal weft or pseudoparenchyma. The outer surface of a non-stromatic perithecium may or may not bear numerous long hairs which when present have taxonomic importance.

In a perithecium, the hymenium either occupies a narrow bit of surface opposite the ostiole (basal portion of the inner surface of the wall) on which asci grow as a small tuft parallel and erect towards the ostiole, or is spread over the entire inner surface of the wall, and the asci then converge radially towards the middle line of the perithecium.

The asci are inserted in the delicate tissue lying inside the perithecium. They fill the inner space of the perithecium or at least the largest part of it, excepting the neck. All the space not occupied by them is filled with branches of the hyphae which grow out from the inner layer of the wall. Some of these branches which are basally attached to hymenium and lie between the asci are termed paraphyses (sing, paraphysis).

Depending on fungi, the asci of perithecium may be evanescent or persistent. In case of evanescent asci the ascus wall with maturity of asci breaks down to form a mass of gelatinous material in which lie embedded the ascospores. This mass of gelatinous material with ascospores is discharged from the perithecium by hydrostatic pressure and remains at the mouth of the ostiole.

The ascospores are then transferred to distant places by the visiting insects which are attracted by the gelatinous material.

Whereas, in case of perithecia bearing persistent asci, the ascospores when mature, are forcibly expelled from them. The ascospores borne in the perithecial ascocarps are one to more than one-celled and are extremely variable in shape and size.

The perithecium of Venturia inaequalis is initiated by the development of a coiled hypha arising within a stroma. The hyphal cells at the periphery of the stroma are Uninucleate, and their walls become thickened, whereas, the cells of the inner hyphae remain thin-walled and multinucleate.

One of these thin-walled cells produces a chain of cells, the ascogonium, each cell of which is bi- to quadri-nucleate.

The apical cell of the ascogonium becomes clavate giving rise to a well-defined trichogyne. Meanwhile, near the developing ascogonium another hyphal tip thickens and becomes lobate, and multinucleate. This is the antheridium. Its growth continues until the lobes contact the trichogyne and become closely applied to it.

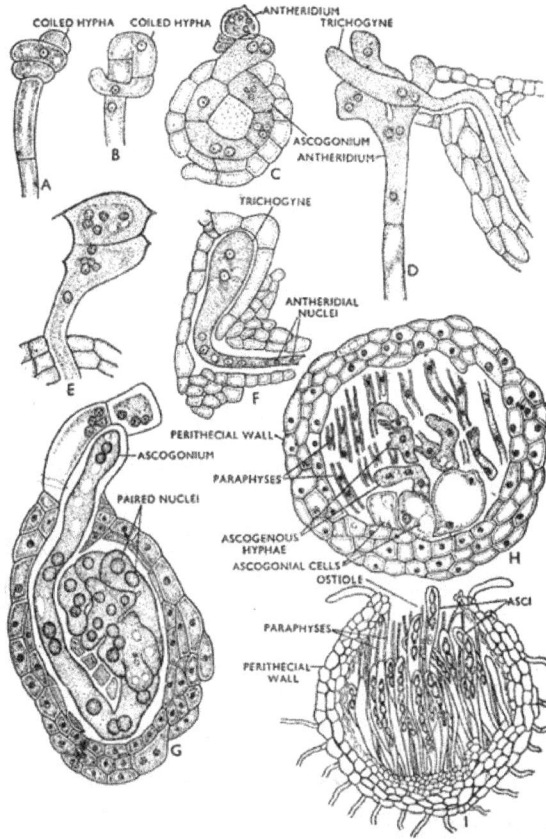

Fig. 212. Development of perithecium in *Venturia inaequalis*. A–B. Stages of coiled hypha development. C. Antheridium and ascogonium developed in stroma. D. The antheridium is in intimate
association with the trichogyne. E. Plasmogamy. F. Trichogyne after nuclear migration, the antheridial nuclei are already at the lower end of the trichogyne. G. Ascogonium with paired nuclei.
H. Section through perithecium with ascogenous hyphae and paraphyses. I. Section of a mature
perithecium showing the nature of perithecial wall, asci with ascospores and paraphyses.

A pore in the adjoining walls then forms, and the antheridial content empties into the
trichogyne. Septa in the ascogonial chain are then, dissolved, whereupon the antheridial nuclei migrate to the ascogonium and become associated in pairs with the ascogonial
nuclei.

Immediately after this, ascogenous hyphae arise as outgrowths from the ascogonium.
Groziers are formed at the tips of the ascogenous hyphae. A binucleate penultimate cell
forms, and the two nuclei fuse promptly to form the diploid stage. The ascus elongates,
and three successive divisions of the diploid nucleus occur producing 8 haploid nuclei,
the first of which is the reduction division.

The cytoplasm is delimited around each of the eight nuclei, and eight uninucleate haploid ascospores are formed. Meanwhile the uninucleate cells of the hyphae multiply to
form the wall of the developing perithecium and the nurse tissue for the developing asci
and paraphyses. Ultimately a perithecium with well-developed perithecial wall bearing
ostiole and containing asci and paraphyses is developed.

Conidium

Conidium, a type of asexual reproductive spore of fungi (kingdom Fungi) usually produced at the tip or side of hyphae (filaments that make up the body of a typical fungus) or on special spore-producing structures called conidiophores. The spores detach when mature.

Conidia which are small spherical, oval, slightly elongated are usually single-celled or with one-septate, are called micro- conidia. Whereas, conidia which are large, usually multi-celled are called macro- conidia. The micro-conidia may be spherical, elliptical pyriform or clavate.

The macro-conidia may be divided into two or more cells by transverse septations and may appear fusiform or long crescent-shaped. Conidia may be thick- or thin-walled having surface smooth or warty and may be extremely variable in shape from globose, oval, cylindrical, clavate, filamentous thread-like, filamentous spirally coiled, to stellate or irregular in shape.

They may be hyaline to bright- or dark-coloured.

Conidia are also named in accordance with their shape or structure, nature of septation and number of cells in a conidium. There may be: one-celled conidia, amerospores; two-celled conidia didymospores; conidia having two or more transverse septa, phragmospores; conidia possessing septa both lengthways and crossways, dictyospores; long, thread- or worm-like conidia are scolecospores; conidia being rolled-up or corkscrew-like cylindrical generally septate are, helicospores; and staurospores, when conidia are star-like having three or more arms.

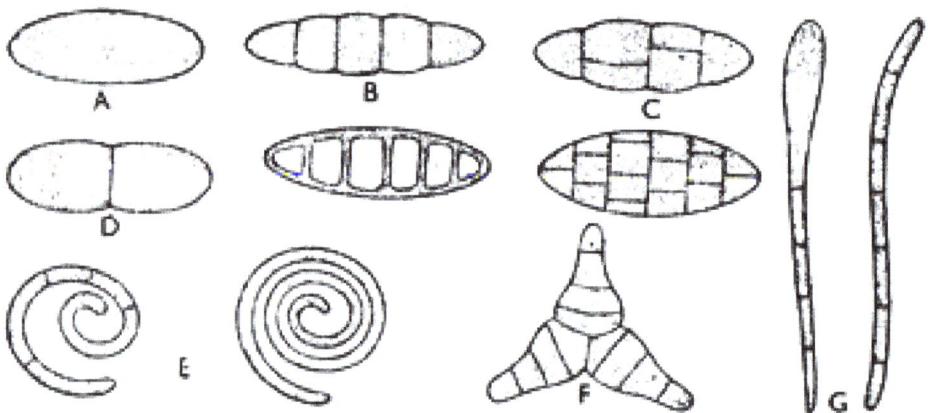

Fig. 302. Grouping of conidia on the basis of colour, shape and septation.

Crozier

In the Ascomycota the cell in which meiosis occurs becomes modified to form a microscopic sporangium-like structure called an ascus (plural: asci).

Many asci arise from hook-like structures called croziers. The picture at right shows the common operculate discomycete Scutellinia scutellata with four small panels to illustrate croziers at the base of developing asci. Croziers are notoriously difficult to see with a microscope and often demand patience on the part of the observer. In this case the asci were stained in the dye Ink Blue dissolved in Shear's Mounting medium so that the cell walls could be seen more clearly. A water solution of Congo Red can also be very useful for demonstrating croziers. Nevertheless you may still have difficulty seeing them without further explanation. The drawings at right, although diagrammatic, illustrate how croziers develop.

Croziers are quite similar in appearance and function to the clamp connections produced by many Basidiomycota. Their function is to pass a nucleus back to a previously-formed cell, thus maintaining the dikaryotic state. But let's examine the process in more detail using the illustration. Drawing Number 1 shows an undifferentiated hypha with two cells, each containing a pair of compatible nuclei. The hypha begins to curve at the top (2) and nuclei move upward to align themselves side-by-side (3). Each nucleus then divides, temporarily giving the crozier four nuclei (4). Two septa are then formed (5), isolating two nuclei in the top of the crozier and one in the cell behind and one in the end cell. These three cells, from the tip back, are usually referred to as the "ultimate", "penultimate", and "antepenultimate" cells of the crozier. The cell walls separating the ultimate and antepenultimate cells then dissolve, allowing the two nuclei to reestablish the dikaryon (6).

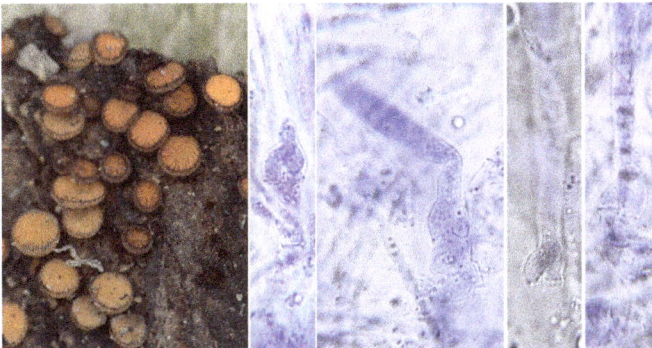

The two nuclei in the penultimate cell fuse to form the diploid (2N) condition (7). At this point the young ascus begins to elongate from the top of the penultimate cell (8). The diploid nucleus undergoes meiosis, leaving the developing ascus with four haploid nuclei (9). There is usually one more cell division to yield eight nuclei, ech of which will become surround by an ascospore wall (10).

Although small and often difficult to demonstrate, croziers are important characteristics used in the identification of many Ascomycota. Their occurrence is sporadic; some species always produce them while others do not. Sometimes two species in the same genus can be separated on their presence and absence. They can be found on all kinds of asci ranging from the large and complex to very simple spherical ones.

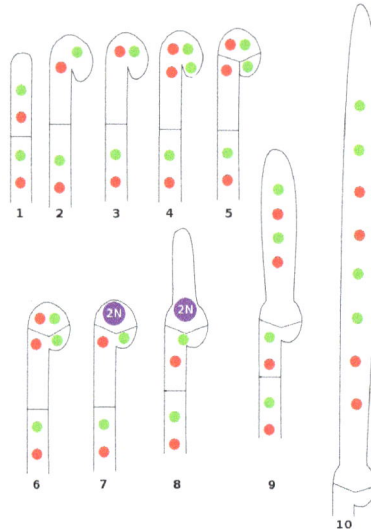

Veil

A veil or velum, in mycology, is one of several structures in fungi, especially the thin membrane that covers the cap and stalk of an immature mushroom. Veils fall into two categories:

Partial veil protective membrane covering gills during development of a fruitbody.

Universal veil a protective membrane that initially surrounds an entire fruitbody.

The critical mission of a mushroom is to disperse its spores. So it is not surprising that many features of the mushroom exist to support this mission. One such feature is the partial veil. The partial veil is a membrane that covers the spore-bearing surface of the mushroom while it is developing. In effect, this encloses the young spore-bearing surface in a small chamber where it's easier for the mushroom to maintain the right humidity and temperature for the developing spore-bearing cells.

Structure

In the immature fruit bodies of some basidiomycete fungi, the partial veil extends from the stem surface to the cap margin and shields the gills during development, and later breaks to expose the mature gills. The presence, absence, or structure of the partial veil is an aid to identification of mushrooms. Some fruit bodies may have both a universal and partial veil, others may have only one or the other, while many lack both types of veils. The partial veil may be membranous or cobwebby, and may have multiple layers. Various adjectives are commonly used to describe the texture of partial veils, such as: *membranous*, like a membrane; *cottony*, where the veil tissue is made of separate fibers that may be easily separated like a cotton ball; *fibrillose*, composed of thin strands and *glutinous*, with a slimy consistency. Some mushrooms have partial veils which are *evanescent*, which are so thin and delicate that they disappear after they rupture, or leave merely a faint trace on the stem known as an *annular zone* or *ring zone*. Others may leave a persistent annulus (ring). Occasionally, the partial veil adheres to the edge of the cap as shreds of tissue, forming an *appendiculate* margin.

A cobweb-like cortina of *Cortinarius claricolor*

The cobweb-like, fragile partial veil of some mushrooms, especially those in the genus *Cortinarius*, are known as *cortinas*. The fibrous threads of the cortina often catch the brown spores as they drop, making them visible as fine brown streaks along the stem. Some species of *Agaricus*, such as *Agaricus arvensis*, have a partial veil that resembles a cogwheel. Mycologists Alexander H. Smith and Harry D. Thiers, in their 1964 monograph on the bolete genus *Suillus*, proposed the term "false veil" to account for those species of *Suillus* that have a "conspicuous cottony roll" of tissue that originates from the cap margin (especially in young specimens) and never becomes integrated with the stem tissue.

The partial veil of some species, like this *Stropharia ambigua*, may form an appendiculate cap margin

Development in Agaricus

After the fruit body of *Armillaria hinnulea* expands, the partial veil remains as a ring on the stem

Species in the genus *Agaricus* have a partial veil that is made of two layers of tissue, although the two layers are not clearly distinct in all species. In the early 20th century, American mycologist George Francis Atkinson investigated the development of the mushroom *Agaricus arvensis* by collecting young mushroom buttons (immature fruit bodies with the veil intact and the cap not yet expanded) and observing their growth in the laboratory. He determined that the partial veil originates from the tissue lying outside the annular cavity (the area containing the delicate developing hymenium and enclosed by the partial veil) and is not clearly separated from the universal veil. It is connected to both the margin of the cap and the surface of the stem. The partial veil increases in size as tension is applied to it from the expansion of the cap and stem. The lower portion of the partial veil (connected to the stem) has a looser texture, and is relatively porous to allow for air exchange. The upper portion of the partial veil (next to the gill cavity) is connected directly with the margin of the cap. It originates partly from fundamental tissue (actively dividing hyphae that comprise the bulk of the cap and stem tissue) and partly by growth from the margin of the cap. The looser portion of the veil is torn off from the surface of the stem as the mushroom expands and grows, and provides the looser lower portion of the duplex veil characteristic of this species and some of the other species of *Agaricus*, like *A. augustus*.

The universal veil is a layer of tissue that completely surrounds the baby mushroom (in some species), making it look like an egg at first (or, more dangerously for the edible-mushroom hunter, a puffball). When the mushroom grows, it breaks out of the universal veil, leaving bits of it attached to various parts of the mushroom.

A universal veil that hangs together like a skin is said to be membranous. The bottom half of a membranous universal veil is often left behind at the base of the mushroom's stalk, forming a sort of a cup that the mushroom seems to be growing out of. This remaining universal veil tissue at the base of the stalk is called a volva, and the projecting part of the volva is called the limb. A volva possessing a limb is called limbate, and the edge of the limb is called the rim of the volva (it is sometimes called the margin, like the edge of the pileus, but I am only using the words in this mutually exclusive senses in this website). A deep volva that is shaped like a cup or sack is called a saccate volva.

Image of Amanita caesarea from Abbé Giacomo Bresadola Iconographia mycologica

A non-limbate volva is clamped tightly to the stalk, and its rim merely separates a small distance or rolls down a bit at the top, like a sock. Such a volva is called peronate, or ocreate. As used here, this term is completely distinct from caligate, which is a similar clamped-to-the-stalk characteristic of the partial veil.

Part of the universal veil can also be left plastered to the cap surface, either as a large central patch or as bits of material that get more and more separated from one another as the cap expands. You will sometimes see these little flakes or bits referred to as "warts" on websites or the more popular literature.

Amanita pantherina var. pantherina from Abbé Giacomo Bresadola
Iconographia mycological

The universal veil can also be powdery. Here, yellow universal veil material (the stuff on the cap is white because it has bleached in the sun) has formed a powdery coating on the base of this mushroom's stalk.

Photo of Amanita frostiana by John Denk

A universal veil may also consist of a layer of slime. In this case, the slime usually coats the stalk and the cap after the mushroom has opened.

A veil of any type that remains on the expanded fruiting body as a volva, annulus or armilla is called persistent. One that is so powdery (or fragile in some other way) that it does not is called evanescent. Fragments of an evanescent veil may remain behind, however, as fragments or powder on the stem, pileus, or hanging from the pileus margin.

Yeast

Yeast, any of about 1,500 species of single-celled fungi, most of which are in the phylum Ascomycota, only a few being Basidiomycota. Yeasts are found worldwide in soils and on plant surfaces and are especially abundant in sugary mediums such as flower nectar and fruits. There are hundreds of economically important varieties of ascomycete yeasts; the types commonly used in the production of bread, beer, and wine are selected strains of *Saccharomyces cerevisiae*. Some yeasts are mild to dangerous pathogens of humans and other animals, especially *Candida albicans*, *Histoplasma*, and *Blastomyces*.

Yeast is a microscopic fungus consisting of single oval cells that reproduce by budding. Some yeasts may form filaments (pseudohyphae, or false hyphae) similar to those formed by molds. There about 1,500 species of yeast currently described, estimated to be only 1% of all fungal species. Yeast is capable of converting sugar into alcohol and carbon dioxide.

Yeast is very common in environments with sugar-rich material such as the skins of fruits and berries (for example grapes, apples or peaches), and exudate from plants (such as plant saps or cacti). Some yeasts are found in association with soil and insects while others including *Candida albicans*, *Rhodotorula rubra*, *Torulopsis* and *Trichosporon cutaneum*, have been found living in between people's toes as part of their skin flora. Yeast is also present in the gut flora of mammals and some insects. Deep-sea environments host an array of yeasts.

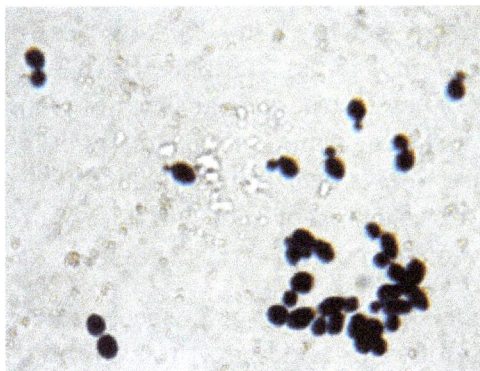

Yeast Cells

Nutrition and Growth

Yeasts are chemoorganotrophs, as they use organic compounds as a source of energy and do not require sunlight to grow. Carbon is obtained mostly from hexose sugars, such as glucose and fructose, or disaccharides such as sucrose and maltose. Some species can metabolize pentose sugars such as ribose, alcohols, and organic acids. Yeast species either require oxygen for aerobic cellular respiration (obligate aerobes) or are anaerobic, but also have aerobic methods of energy production (facultative anaerobes). Unlike bacteria, no known yeast species grow only anaerobically (obligate anaerobes). Most yeasts grow best in a neutral or slightly acidic pH environment.

Yeasts vary in regard to the temperature range in which they grow best. For example, *Leucosporidium frigidum* grows at −2 to 20 °C (28 to 68 °F), *Saccharomyces telluris* at 5 to 35 °C (41 to 95 °F), and *Candida slooffi* at 28 to 45 °C (82 to 113 °F). The cells can survive freezing under certain conditions, with viability decreasing over time.

In general, yeasts are grown in the laboratory on solid growth media or in liquid broths. Common media used for the cultivation of yeasts include potato dextrose agar or potato dextrose broth, Wallerstein Laboratories nutrient agar, yeast peptone dextrose agar, and yeast mould agar or broth. Home brewers who cultivate yeast frequently use dried malt extract and agar as a solid growth medium. The antibiotic cycloheximide is sometimes added to yeast growth media to inhibit the growth of *Saccharomyces* yeasts and select for wild/indigenous yeast species. This will change the yeast process.

The appearance of a white, thready yeast, commonly known as kahm yeast, is often a byproduct of the lactofermentation (or pickling) of certain vegetables, usually the result of exposure to air. Although harmless, it can give pickled vegetables a bad flavor and must be removed regularly during fermentation.

Ecology

Yeasts are very common in the environment, and are often isolated from sugar-rich materials. Examples include naturally occurring yeasts on the skins of fruits and berries (such as grapes, apples, or peaches), and exudates from plants (such as plant saps or cacti). Some yeasts are found in association with soil and insects. The ecological function and biodiversity of yeasts are relatively unknown compared to those of other microorganisms. Yeasts, including *Candida albicans*, *Rhodotorula rubra*, *Torulopsis* and *Trichosporon cutaneum*, have been found living in between people's toes as part of their skin flora. Yeasts are also present in the gut flora of mammals and some insects and even deep-sea environments host an array of yeasts.

An Indian study of seven bee species and 9 plant species found 45 species from 16 genera colonise the nectaries of flowers and honey stomachs of bees. Most were members of the genus *Candida*; the most common species in honey stomachs was *Dekkera intermedia* and in flower nectaries, *Candida blankii*. Yeast colonising nectaries of the stinking hellebore have been found to raise the temperature of the flower, which may aid in attracting pollinators by increasing the evaporation of volatile organic compounds. A black yeast has been recorded as a partner in a complex relationship between ants, their mutualistic fungus, a fungal parasite of the fungus and a bacterium that kills the parasite. The yeast has a negative effect on the bacteria that normally produce antibiotics to kill the parasite, so may affect the ants' health by allowing the parasite to spread.

Certain strains of some species of yeasts produce proteins called yeast killer toxins that allow them to eliminate competing strains. This can cause problems for winemaking but could potentially also be used to advantage by using killer toxin-producing strains to make the wine. Yeast killer toxins may also have medical applications in treating yeast infections.

Haploid and Diploid Cells

To understand either type of reproduction in yeast, one must also understand that yeast exist with both haploid and diploid cells. Haploid cells have only one set of chromosomes, whereas diploid cells have a set of chromosomes from each parent cell. For example, humans are always diploid -- having 46 chromosomes, with 23 coming from each parent, which then end up in pairs. But human sex cells, the sperm and egg, are haploid with only one set of chromosomes each that then combine when the sperm fertilizes the egg. But yeast cells can be either; haploid and diploid cells may coexist. Haploid cells come in two "genders" also, similarly to humans, although diploid cells do not have a gender.

Asexual Reproduction in Yeast

The asexual form of reproduction in yeast is called fission, or sometimes "budding." Budding is exactly what it sounds like. The parent cell begins to divide to form a new cell, which is the "daughter" cell, by splitting its nucleus and copying the contents, thus migrating the new nucleus into the daughter cell. The process is basically standard mitosis (cell division). The newly created cell is an exact copy of the parent cell; it can be either diploid or haploid.

Sexual Reproduction in Yeast

Only haploid yeast cells are able to conduct sexual reproduction. When they do, the haploid cells are usually not the same gender. Before joining with the opposite type of haploid yeast cell, each cell undergoes a process called shmooing in which it becomes longer and thinner in preparation for the joining. The shmooing cells then fuse and join their nuclei together to create a diploid. The new diploid then begins to bud and form a colony of diploid yeast cells.

Sexual reproduction continues with the diploid cell returning to a haploid state when the environmental conditions become hostile. Cells that already exist as haploids tend to die off when environmental conditions are unfavorable, but diploid cells undergo meiosis and subsequently divide into four daughter haploid spores in poor environmental conditions. Spores are inert cells that are able to tolerate adverse conditions much better than active cells. When the environmental conditions become favorable again, the spores begin to resume normal lifecycle activities, either continuing to exist as haploids or joining with other haploids to create new diploid colonies.

Environmental Conditions for Yeast Reproduction

In order for yeast to exist and reproduce outside of the spore state, the cells generally require a lukewarm, moist environment with some access to sugar. Yeast reproduction in bread dough is what causes bread to rise, and the cells are then killed when the dough has been baked.

How Yeast Works

Through the process of fermentation, yeast converts sugars into carbon dioxide and alcohol. These two byproducts make yeast an extremely useful tool in food production.

Carbon dioxide is what gives alcoholic beverages such as beer and champagne their characteristic bubbles and is also responsible for rising bread. As yeast begins to metabolize sugars in bread dough, the carbon dioxide gas is trapped within the gluten strands, creating bubbles, and causing a leavening action. In beverages, the carbon dioxide is trapped within the liquid by the pressure of the sealed container. When the container is opened, the pressure is released and the carbon dioxide begins to release or bubble.

Alcohol, the other byproduct of yeast fermentation, is also produced during the bread making process but evaporates as the bread bakes. When making alcoholic beverages, yeast is allowed to ferment for a much longer period of time, allowing it to produce more alcohol.

Culinary Uses for Yeast

- Beer: Yeast is added to malted grains and allowed to ferment in order to produce alcohol. The type of yeast used will affect the type and flavor of beer produced. S. cerevisiae, also known as a "top fermenting" or "top cropping" yeast, ferments at a higher temperature and produces sweet or fruity beers. Bottom croppers, such as *Saccharomyces pastorianus*, ferment at lower temperatures and are used to make lagers.

- Wine: Yeast is naturally present on the skins of grapes and can be used to naturally ferment grape juice into wine. Despite the naturally present yeast, most wines today have pure culture (usually S. cerevisiae) added to them to produce a more consistent and controllable result. There are many different strains of S. cerevisiae and each will produce a unique flavor characteristic in a finished wine.

- Bread: Records of using yeast as a leavening agent date back to the ancient Egyptians, though the form of yeast used has changed over time. Many varieties are available for use in bread making, such as fresh yeast cakes, bakers yeast, active dry, instant, or yeast starter.

Bioremediation

Some yeasts can find potential application in the field of bioremediation. One such yeast, *Yarrowia lipolytica*, is known to degrade palm oil mill effluent, TNT (an explosive material), and other hydrocarbons, such as alkanes, fatty acids, fats and oils. It can also tolerate high concentrations of salt and heavy metals, and is being investigated for its potential as a heavy metal biosorbent. *Saccharomyces cerevisiae* has potential to bioremediate toxic pollutants like arsenic from industrial effluent. Bronze statues are known to be degraded by certain species of yeast. Different yeasts from Brazilian gold mines bioaccumulate free and complexed silver ions.

Industrial Ethanol Production

The ability of yeast to convert sugar into ethanol has been harnessed by the biotechnology industry to produce ethanol fuel. The process starts by milling a feedstock, such as sugar cane, field corn, or other cereal grains, and then adding dilute sulfuric acid, or fungal alpha amylase enzymes, to break down the starches into complex sugars. A glucoamylase is then added to break the complex sugars down into simple sugars. After this, yeasts are added to convert the simple sugars to ethanol, which is then distilled off to obtain ethanol up to 96% in purity.

Saccharomyces yeasts have been genetically engineered to ferment xylose, one of the major fermentable sugars present in cellulosic biomasses, such as agriculture residues, paper wastes, and wood chips. Such a development means ethanol can be efficiently produced from more inexpensive feedstocks, making cellulosic ethanol fuel a more competitively priced alternative to gasoline fuels.

Nonalcoholic Beverages

A *kombucha* culture fermenting in a jar

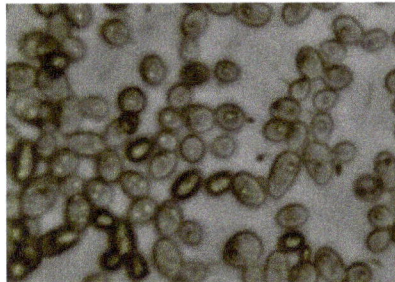

Yeast and bacteria in kombucha at 400×

A number of sweet carbonated beverages can be produced by the same methods as beer, except the fermentation is stopped sooner, producing carbon dioxide, but only trace amounts of alcohol, leaving a significant amount of residual sugar in the drink.

- Root beer, originally made by Native Americans, commercialized in the United States by Charles Elmer Hires and especially popular during Prohibition

- Kvass, a fermented drink made from rye, popular in Eastern Europe. It has a recognizable, but low alcoholic content.

- Kombucha, a fermented sweetened tea. Yeast in symbiosis with acetic acid bacteria is used in its preparation. Species of yeasts found in the tea can vary, and may include: *Brettanomyces bruxellensis, Candida stellata, Schizosaccharomyces pombe, Torulaspora delbrueckii* and *Zygosaccharomyces bailii*. Also popular in Eastern Europe and some former Soviet republics under the name *chajnyj grib* (Russian: Чайный гриб), which means "tea mushroom".

- Kefir and kumis are made by fermenting milk with yeast and bacteria.

- Mauby (Spanish: *mabí*), made by fermenting sugar with the wild yeasts naturally present on the bark of the *Colubrina elliptica* tree, popular in the Caribbean

Nutritional Supplements

Yeast is used in nutritional supplements, especially those marketed to vegans. It is often referred to as "nutritional yeast" when sold as a dietary supplement. Nutritional yeast is a deactivated yeast, usually *S. cerevisiae*. It is naturally low in fat and sodium as well as an excellent source of protein and vitamins, especially most B-complex vitamins (contrary to some claims, it contains little or no vitamin B_{12}), as well as other minerals and cofactors required for growth. Some brands of nutritional yeast, though not all, are fortified with vitamin B_{12}, which is produced separately by bacteria.

In 1920, the Fleischmann Yeast Company began to promote yeast cakes in a "Yeast for Health" campaign. They initially emphasized yeast as a source of vitamins, good for skin and digestion. Their later advertising claimed a much broader range of health benefits, and was censured as misleading by the Federal Trade Commission. The fad for yeast cakes lasted until the late 1930s.

Nutritional yeast has a nutty, cheesy flavor and is often used as an ingredient in cheese substitutes. Another popular use is as a topping for popcorn. It can also be used in mashed and fried potatoes, as well as in scrambled eggs. It comes in the form of flakes, or as a yellow powder similar in texture to cornmeal. In Australia, it is sometimes sold as "savoury yeast flakes". Though "nutritional yeast" usually refers to commercial products, inadequately fed prisoners have used "home-grown" yeast to prevent vitamin deficiency.

Probiotics

Some probiotic supplements use the yeast *S. boulardii* to maintain and restore the natural flora in the gastrointestinal tract. *S. boulardii* has been shown to reduce the symptoms of acute diarrhea, reduce the chance of infection by *Clostridium difficile* (often identified simply as C. difficile or C. diff), reduce bowel movements in diarrhea-predominant IBS patients, and reduce the incidence of antibiotic-, traveler's-, and HIV/AIDS-associated diarrheas.

Aquarium Hobby

Yeast is often used by aquarium hobbyists to generate carbon dioxide (CO_2) to nourish plants in planted aquaria. CO_2 levels from yeast are more difficult to regulate than those from pressurized CO_2 systems. However, the low cost of yeast makes it a widely used alternative.

Yeast Extract

Marmite and Vegemite, products made
from yeast extract

Marmite and Vegemite are dark in
colour

Yeast extract is the common name for various forms of processed yeast products that are used as food additives or flavours. They are often used in the same way that monosodium glutamate (MSG) is used and, like MSG, often contain free glutamic acid. The general method for making yeast extract for food products such as Vegemite and Marmite on a commercial scale is to add salt to a suspension of yeast, making the solution hypertonic, which leads to the cells' shrivelling up. This triggers autolysis, wherein the yeast's digestive enzymes break their own proteins down into simpler compounds, a process of self-destruction. The dying yeast cells are then heated to complete their breakdown, after which the husks (yeast with thick cell walls that would give poor texture) are separated. Yeast autolysates are used in Vegemite and Promite (Australia); Marmite, Bovril and Oxo (the United Kingdom, Republic of Ireland and South Africa); and Cenovis (Switzerland).

Scientific Research

Several yeasts, in particular *S. cerevisiae*, have been widely used in genetics and cell biology, largely because *S. cerevisiae* is a simple eukaryotic cell, serving as a model for all eukaryotes, including humans, for the study of fundamental cellular processes such as the cell cycle, DNA replication, recombination, cell division, and metabolism. Also, yeasts are easily manipulated and cultured in the laboratory, which has allowed for the development of powerful standard techniques, such as yeast two-hybrid, synthetic genetic array analysis, and tetrad analysis. Many proteins important in human biology were first discovered by studying their homologues in yeast; these proteins include cell cycle proteins, signaling proteins, and protein-processing enzymes.

On 24 April 1996, *S. cerevisiae* was announced to be the first eukaryote to have its genome, consisting of 12 million base pairs, fully sequenced as part of the Genome Project. At the time, it was the most complex organism to have its full genome sequenced, and the work seven years and the involvement of more than 100 laboratories to accomplish. The second yeast species to have its genome sequenced was *Schizosaccharomyces pombe*, which was completed in 2002. It was the sixth eukaryotic genome sequenced and consists of 13.8 million base pairs. As of 2014, over 50 yeast species have had their genomes sequenced and published.

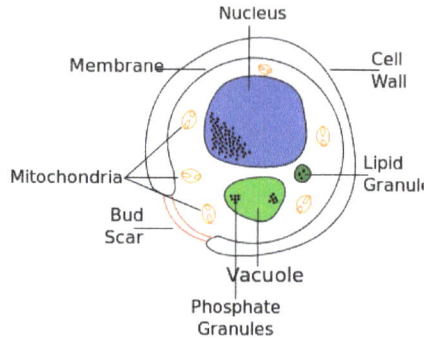

Diagram showing a yeast cell

Genetically Engineered Biofactories

Various yeast species have been genetically engineered to efficiently produce various drugs, a technique called metabolic engineering. *S. cerevisiae* is easy to genetically engineer; its physiology, metabolism and genetics are well known, and it is amenable for use in harsh industrial conditions. A wide variety of chemical in different classes can be produced by engineered yeast, including phenolics, isoprenoids, alkaloids, and polyketides. About 20% of biopharmaceuticals are produced in *S. cerevisiae*, including insulin, vaccines for hepatitis, and human serum albumin.

Pathogenic Yeasts

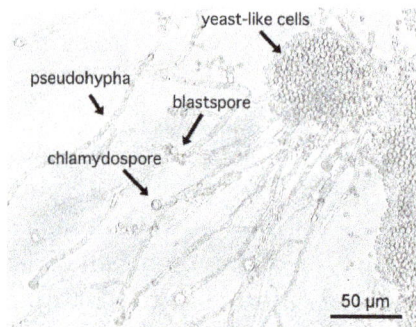

A photomicrograph of *Candida albicans* showing
hyphal outgrowth and other morphological characteristics

Some species of yeast are opportunistic pathogens that can cause infection in people with compromised immune systems. *Cryptococcus neoformans* and *Cryptococcus gattii* are significant pathogens of immunocompromised people. They are the species primarily responsible for cryptococcosis, a fungal disease that occurs in about one million HIV/AIDS patients, causing over 600,000 deaths annually. The cells of these yeast are surrounded by a rigid polysaccharide capsule, which helps to prevent them from being recognised and engulfed by white blood cells in the human body.

Yeasts of the genus *Candida*, another group of opportunistic pathogens, cause oral and vaginal infections in humans, known as candidiasis. *Candida* is commonly found as a commensal yeast in the mucous membranes of humans and other warm-blooded animals. However, sometimes these same strains can become pathogenic. The yeast cells sprout a hyphal outgrowth, which locally penetrates the mucosal membrane, causing irritation and shedding of the tissues. The pathogenic yeasts of candidiasis in probable descending order of virulence for humans are: *C. albicans*, *C. tropicalis*, *C. stellatoidea*, *C. glabrata*, *C. krusei*, *C. parapsilosis*, *C. guilliermondii*, *C. viswanathii*, *C. lusitaniae*, and *Rhodotorula mucilaginosa*. *Candida glabrata* is the second most common *Candida* pathogen after *C. albicans*, causing infections of the urogenital tract, and of the bloodstream (candidemia).

Food Spoilage

Yeasts are able to grow in foods with a low pH (5.0 or lower) and in the presence of sugars, organic acids, and other easily metabolized carbon sources. During their growth, yeasts metabolize some food components and produce metabolic end products. This causes the physical, chemical, and sensible properties of a food to change, and the food is spoiled. The growth of yeast within food products is often seen on their surfaces, as in cheeses or meats, or by the fermentation of sugars in beverages, such as juices, and semiliquid products, such as syrups and jams. The yeast of the genus *Zygosaccharomyces* have had a long history as spoilage yeasts within the food industry. This is mainly because these species can grow in the presence of high sucrose, ethanol, acetic acid, sorbic acid, benzoic acid, and sulphur dioxide concentrations, representing some of the commonly used food preservation methods. Methylene blue is used to test for the presence of live yeast cells. In oenology, the major spoilage yeast is *Brettanomyces bruxellensis*.

Microfungi

Microscopic fungi are eukaryotic, heterotrophic microorganisms that fail to show any cellular differentiation into true tissues like root, stem or leaf and in which vascular system is absent.

Microfungi comprise a loosely defined artificial group of Fungi and fungal-like organisms that include such things as bread molds, plant pathogens, powdery mildews, rusts, slime molds, and water molds. In general, these fungi are difficult or impossible to see with the unaided eye. A taxonomical classification of microfungi suggests the group contains 4468 genera and 55,989 species.

Microfungi are ubiquitous throughout the world and some cause major economic impacts as pathogens of animals, plants, and other fungi. Many microfungi are harmless saprobes, breaking down large complex chemical structures such as lignin found in wood into usable simple compounds. Despite their importance, little is known about the diversity, distribution, ecology, or host relationships of microfungi throughout the United States.

Harmful microfungi

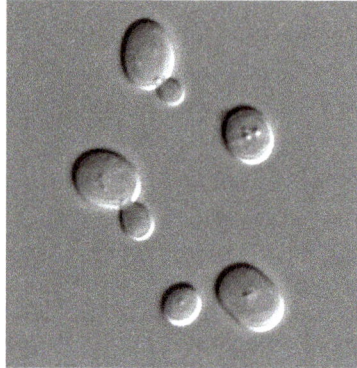

Yeast of the species *Saccharomyces cerevisiae*

Microfungi can also be harmful, causing diseases of plants, animals and humans with varying degrees of severity and economic impact. The irritating human skin disease known as athlete's foot or tinea pedis is caused by species of the microfungal genus *Trichophyton*. Microfungi may cause diseases of crops and trees which range in severity from mild to disastrous, and in economic importance from beneficial to seriously costly. The mould *Botrytis cinerea* can cause spoilage of crops including grapes, but is also responsible for the "noble rot", which concentrates sugars in the grapes used to make the intensely sweet and concentrated Sauternes dessert wines from the Bordeaux region of France. The potato famine in Ireland during the mid-to-late 19th century was caused by a fungus called *Phytophthora infestans* that rotted the potato crops for several years. Dutch elm disease, which has ravaged elms across Europe and North America in the last 50 years, is caused by the microfungi of the genus *Ophiostoma*. Rice blast, a devastating fungal disease of cereals including rice, wheat and millet, is caused by the phytopathogenic Ascomycete fungus *Magnaporthe grisea*. In the built environment, the toxic fungus *Stachybotrys chartarum* causes damage to damp walls and furnishings, and may be responsible for sick building syndrome.

Types of epidermal microfungal infections are:

- Yeast infection

- Athlete's foot

- Mycosis

- Tinea

- Candida

Diversity

Within the United States, approximately 13,000 species of microfungi on plants or plant products are thought to exist. Specimens of microfungi are housed in the U.S.

National Fungus Collections and other institutions that serve as reservoirs of information and documentation about the nation's natural heritage. Based on the number of species reported in the literature and those represented in the collections; the number of microfungi known in the United States is estimated at 29,000 species. In areas of the world where fungi have been well studied, the ratio of vascular plants to fungi is about 6 to 1. This suggests that there may be as many as 120,000 species of fungi within the United States and 1.5 million worldwide.

Mold

Mold (or mould) is a term used to refer to fungi that grow in the form of multicellular thread-like structures called hyphae. Fungi that exist as single cells are called yeasts. Some molds and yeasts cause disease or food spoilage, others play an important role in biodegradation or in the production of various foods, beverages, antibiotics and enzymes.

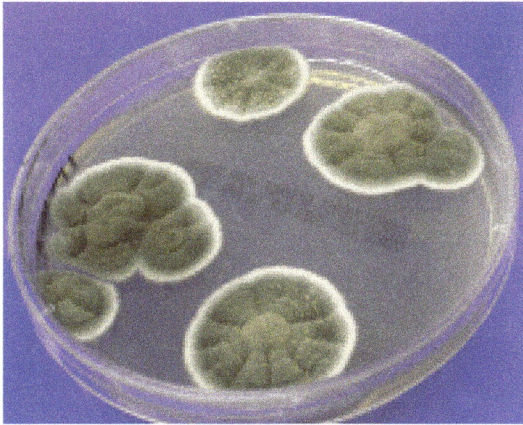

Colonies of Penicillium mold

Mold is also found in damp building materials where it often appears like stains and comes in a variety of colours. A must smell is an indication of microbial growth even when there is no visible growth. If mold is allowed to grow in homes or offices it can contribute to poor indoor air quality. Some molds such as the Dry Rot Fungus, Serpula lacrymans, are highly destructive.

Mold growth requires moisture. The sources of moisture could be Washing, cooking, air humidifiers, condensation or leaks from plumbing or from the outside. Poor ventilation contributes to higher humidity levels and leads to condensation, which also allows mold growth.

Molds release small "spores" into the air. So, when mold grows indoors, the number of mold spores and fragments is usually higher indoors than it is outdoors. These spores are small enough that people can actually inhale them deep into the lungs. Inhalation of spores poses risk of developing respiratory problems. With the exception of winter months mold spores are always present outdoors.

Common Molds

Alternaria

One of the four most common allergenic molds, grows on plants and plant material, easily made air-borne, released by the wind during dry periods

Aspergillus

One of the four most common allergenic molds, common soil fungus, released by the wind during dry periods, will grow on almost any substrate, frequently found on damp hay, grain, sausage, and fruits, commonly cultured from houses, especially basements, crawl spaces, and bedding

Botrytis

Damaged or dead plants, dead leaves, and occasionally prunings and fallen fruit, also can attack flowering plants, especially petals, and fruits such as citrus, kiwi, and grapes

Candida

Normally found on the skin and on mucosal surfaces of humans, including mouth, vagina, and intestinal tract

Cephalosporium

Common soil inhabitant, also isolated from dust in textile plants

Cephalothecium

One of many names for a group of fungi that causes fruit to rot, especially apples.

Chaetomium

Normally found in soil, also grows well on damp paper, fabric, and straw

Curvularia

Soil-borne organism, also grows on plants and plant material, spores are easily dispersed by mowing the grass, released by the wind during dry periods .

Epicoccum

Normally a soil organism, released by the wind during dry periods, often found on decaying plant and vegetable material, plant leaves, uncooked fruit

Fusarium

Common field fungus, parasite on green plants such as peas, beans, cotton, tomatoes, com, sweet potatoes, melons and rice, also saprophytic on decaying plants, loosened during wet periods and dispersed by raindrops

Gliocladium

Normal soil inhabitant, also grows on decomposing plant debris, damp canvas, and occasionally on finished wood products such as paper or cardboard

Helminthosporium

Common field fungus, parasite on cereal grain plants such as com, wheat, oats, and rye, released by the wind during dry periods

Hormodendrum

One of the four most common allergenic molds, found on decomposing plants, leather, rubber, cloth, paper and wood products, released outdoors in great numbers after rains and damp weather

Monilia

Soil borne, also frequently grows on bread and pastries Normal soil, soil around barns and barnyards

Mucor

Normal soil, soil around barns and barnyards

Mycogone species

Found in soil, casing, and straw

Nigrospora

Living grasses but also present on dead ones; easily isolated from dead lawn grass in the autumn, released by the wind during dry periods, also causes disease of certain fruits and vegetables

Paecilomyces

Common soil inhabitant, also grows on damp paper and decaying vegetable material

Penicillium

One of the four most common allergenic molds, soil inhabitant, grows readily on fruits, breads, cheese and other foods, commonly cultured from houses, especially basements, crawl spaces, and bedding

Phoma

Paper products such as books and magazines, also certain paints and green plants, outdoors loosened during wet periods and dispersed by raindrops

Pullularia

Normally in soil, also decaying vegetation, plants and chalking compounds, outdoors loosened during wet periods and dispersed by raindrops

Rhizopus

Grows readily on bread, cured meats and root vegetables indoors, on a variety of plants outdoors

Rhodotorula

A yeast, splash-dispersed, prominent at night and in wet weather, high during fall rains, especially where fruit and berry crops are grown, also often indoors

Saccharomyces

Brewer's yeast, widespread in nature, occurring in soil and on plants

Spondylocladium

Decaying wood, potatoes and other plant material, released by the wind during dry periods

Stemphylium

Damp paper, canvas and cotton fabric, decaying plant material

Trichoderma

Decaying wood, pine stumps, damp cotton and wool, damp basements, outdoors loosened during wet periods and dispersed by raindrops

Trichophyton

Found as a fungal infection of the skin, primarily on the foot

Growth in Buildings and Homes

Moldy housecorner from outside and inside

Mold growth in buildings can lead to a variety of health problems as microscopic airborne reproductive spores, analogous to tree pollen, are inhaled by building occupants. High quantities of indoor airborne spores as compared to exterior conditions are strongly suggestive of indoor mold growth. Determination of airborne spore counts is accomplished by way of an air sample, in which a specialized pump with a known flow rate is operated for a known period of time. Conducive to scientific methodology, air samples should be drawn from the affected area, a control area, and the exterior.

The air sampler pump draws in air and deposits microscopic airborne particles on a culture medium. The medium is cultured in a laboratory and the fungal genus and species are determined by visual microscopic observation. Laboratory results also quantify fungal growth by way of a spore count for comparison among samples. The pump operation time was recorded and when multiplied by the operation time results in a specific volume of air obtained. Although a small volume of air is actually analyzed, common laboratory reporting techniques extrapolate the spore count data to equate the amount of spores that would be present in a cubic meter of air.

Various practices can be followed to mitigate mold issues in buildings, the most important of which is to reduce moisture levels that can facilitate mold growth. Properly functioning air conditioning (AC) units are essential to controlling levels of indoor airborne fungal spores. Air filtration reduces the number of spores available for germination, especially when a High Efficiency Particulate Air (HEPA) filter is used. A properly functioning AC unit also reduces the relative humidity, or the moisture inherent in the air. The United States Environmental Protection Agency (EPA) currently recommends that relative humidity be maintained between 30% to 50% to preempt mold growth. Considering that fungal growth requires cellulose, plant fiber, as a food source, using building materials that do not contain cellulose is an effective method of preventing fungal growth.

Eliminating the moisture source is the first step at fungal remediation. Removal of affected materials may also be necessary for remediation, if materials are easily replaceable and not part of the load-bearing structure. Professional drying of concealed wall cavities and enclosed spaces such as cabinet toekick spaces may be required. Post-remediation verification of moisture content and fungal growth is required for successful remediation. Many contractors perform post-remediation verification themselves, but property owners may benefit from independent verification.

Use of Molds for Medicine Production:

There are several species of molds which are used for production of penicillin such as *penicillium crysogenum, penicillium rubens* etc. There are several other drugs/medications which are produced from mold such as statin group of medicines (such as lovastatin from *Aspergillus terreus*) which are used for lowering blood cholesterol level and very commonly used medicines. Immune suppressant medicine cyclosporine (this is used for suppressing immunity after organ transplantation, so that body do not reject the transplanted organ such as kidney), is produced from a mold known as *tolypocladium inflatum.*

Use of Molds in Food Production:

Molds are commonly used for food production, throughout the world in various methods. Molds are being used for centuries for producing various foods in many countries across the globe. Molds are used for fermenting soybeans and wheat to produce soybean paste and soy sauce. Molds are used for breaking down starch fond in rice, barley, sweet potato etc. to produce spirit/alcohol, a process known as *saccharification.*

Molds are also used for production of various types of foods, for example:

- Mold *geotrichum candidum* and some *penicillium spp.* used for production of cheese.

- Mold *ustilago maydis* used for prodcing tortilla based foods.

- *Rhizomucor spp.* used for making vegetarian cheese.

- Many types of molds are used as flavoring agents.

- Yeasts and molds are used for many other food preparations to add softness and to add special taste to various foods.

Harmful Effects of Molds

The type and severity of health effects that result from molds exposure is widely variable among different locations, from person to person and over time.

Although difficult to predict, exposure to molds growing indoors is most often associated with the following allergy symptoms:

- Nasal and sinus congestion
- Cough/sore throat
- Chest tightness
- Dyspnea (breathing difficulty)
- Asthma (or exacerbation of it)
- Epistaxis (nosebleed)
- Upper respiratory tract infections
- Headache
- Skin and eye irritation

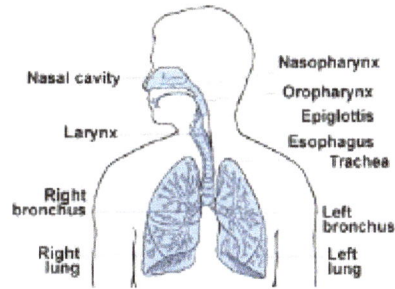

Indoor molds exposure leads mostly to upper respiratory tract symptoms

Long-term exposure to indoor molds is certainly unhealthy to anyone, but some groups will develop more severe symptoms sooner than others, including:

- Infants and children
- Elderly people
- Individuals with respiratory conditions, allergies and/or asthma
- Immunocompromised patients

Some indoor molds are capable of producing extremely potent toxins (mycotoxins) that are lipid-soluble and readily absorbed by the intestinal lining, airways, and skin. These agents, usually contained in the fungal spores, have toxic effects ranging from short-term irritation to immunosuppression and cancer.

More severe symptoms that could result from continuous human exposure to indoor mycotoxigenic molds include:

- Cancer (aflatoxin best characterized as potential human carcinogen)
- Hypersensitivity pneumonitis/pulmonary fibrosis
- Pulmonary injury/hemosiderosis (bleeding)

- Neurotoxicity

- Hematologic and immunologic disorders

- Hepatic, endocrine and/or renal toxicities

- Pregnancy, gastrointestinal and/or cardiac conditions

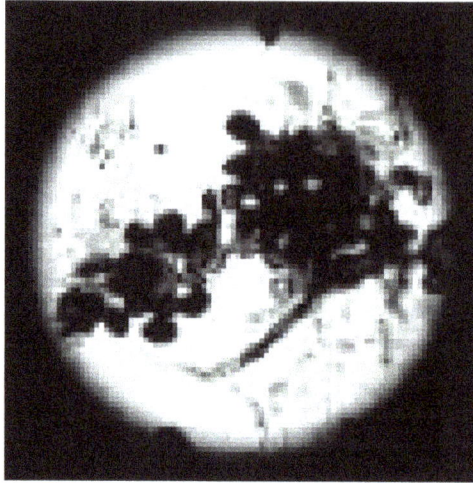

It is important to notice that the clinical relevance of mycotoxins under realistic airborne exposure levels is not fully established. Further, some or much of the supporting evidence for these other health effects is based on case studies rather than controlled studies, studies that have not yet been reproduced or involve symptoms that are subjective.

References

- Kirk PM, Cannon PF, Minter DW, Stalpers JA (2008). Dictionary of the Fungi (10th ed.). Wallingford, UK: CABI. p. 500. ISBN 978-0-85199-826-8

- González Techera A, Jubany S, Carrau FM, Gaggero C (2001). "Differentiation of industrial wine yeast strains using microsatellite markers". Letters in Applied Microbiology. 33 (1): 71–75. doi:10.1046/j.1472-765X.2001.00946.x. PMID 11442819

- Atkinson GF. (1914). "The development of Agaricus arvensis and A. comtulus". American Journal of Botany. 1 (1): 3–22. doi:10.2307/2434957. JSTOR 2434957

- Roody WC. (2003). Mushrooms of West Virginia and the Central Appalachians. Lexington, Kentucky: University Press of Kentucky. p. 499. ISBN 0-8131-9039-8

- "Research enables yeast supplier to expands options" (PDF). Archived from the original (PDF) on 21 September 2006. Retrieved 10 January 2007

- Hoffman CS, Wood V, Fantes PA (October 2015). "An Ancient Yeast for Young Geneticists: A Primer on the Schizosaccharomyces pombe Model System". Genetics. 201 (2): 403–23. doi:10.1534/genetics.115.181503. PMC 4596657. PMID 26447128

- Walker K, Skelton H, Smith K (2002). "Cutaneous lesions showing giant yeast forms of Blastomyces dermatitidis". Journal of Cutaneous Pathology. 29 (10): 616–618. doi:10.1034/j.1600-0560.2002.291009.x. PMID 12453301

- Bessette AR, Bessette A, Roody WC (2000). North American Boletes: A Color Guide to the Fleshy Pored Mushrooms. Syracuse, New York: Syracuse University Press. p. 7. ISBN 0-8156-0588-9

- Ross JP (September 1997). "Going wild: wild yeast in winemaking". Wines & Vines. Archived from the original on 5 May 2005. Retrieved 15 January 2012

- McBryde C, Gardner JM, de Barros Lopes M, Jiranek V (2006). "Generation of novel wine yeast strains by adaptive evolution". American Journal of Enology and Viticulture. 57 (4): 423–430

- Brat D, Boles E, Wiedemann B (2009). "Functional expression of a bacterial xylose isomerase in Saccharomyces cerevisiae". Applied and Environmental Microbiology. 75 (8): 2304–2311. doi:10.1128/AEM.02522-08. PMC 2675233. PMID 19218403

Mycorrhiza and their Types

Mycorrhiza refers to a symbiotic association between the roots of a vascular host plant and a fungus. In such an association, a fungus colonizes a host plant's root tissues. The aim of this chapter is to explore the different types of mycorrhizal associations, such as Arbuscular mycorrhiza, Ectomycorrhiza, Ericoid mycorrhiza, Orchid mycorrhiza, etc.

Mycorrhiza

Mycorrhiza, also spelled Mycorhiza, an intimate association between the branched, tubular filaments (hyphae) of a fungus (kingdom Fungi) and the roots of higher plants. The association is usually of mutual benefit (symbiotic): a delicate balance between host plant and symbiont results in enhanced nutritional support for each member. The establishment and growth of certain plants (*e.g.,* citrus, orchids, pines) is dependent on mycorrhiza; other plants survive but do not flourish without their fungal symbionts. The two main types of mycorrhiza are endotrophic, in which the fungus invades the hosts' roots (*e.g.,* orchids), and ectotrophic, in which the fungus forms a mantle around the smaller roots (*e.g.,* pines). Exploitation of these natural associations can benefit forestry, horticulture, and other plant industries.

Benefits of Mycorrhizal Inoculation

In situations where naturally occurring mycorrhizal associations are infrequent or ineffective an appropriate inoculant can benefit new or established plants by:

- Reducing transplant shock
- Extending the growing season - plants grow larger, flower earlier and produce higher yields
- Providing protection from attack by soil-borne pathogens
- Buffering against toxic levels of trace elements on contaminated land
- Increasing the ability to tolerate environmental stresses.

Commercially available products often contain both ectomycorrhizal and endomycorrhizal species and allow the plant to select the appropriate species for its needs. Not all plants benefit equally from the symbiosis; plant responses to interaction with mycorrhizal fungi vary considerably and management practices such as tillage and crop rotation may adversely affect mycorrhizal structures. Mycorrhizae are especially beneficial for plants in degraded or infertile soils and generally coarse-rooted plants benefit more than fine-rooted plants.

Mycorrhiza also benefit the soil and the wider ecosystem in the following ways:

- Improving soil structure, particularly in manufactured or degraded soils

- Allowing survival of many kinds of seedling that would otherwise never compete, in effect increasing plant diversity

- Below-ground diversity of fungi is one of the major contributing factors to the above-ground plant biodiversity in ecosystems.

Arbuscular Mycorrhiza

Arbuscular (AM) endomycorrhizas are the most common type of mycorrhizal association, and were probably the first to evolve; the fungi are members of the Glomeromycota. In other textbooks you may find these fungi placed in the Order Glomales and Phylum Zygomycota but this is incorrect. The AM fungi are obligate biotrophs, and they are associated with roots of about 80% of plant species (that's equivalent to about two-thirds of all land plants, or around 90% of all vascular plants), including many crop plants. The AM association is endotrophic, and has previously been referred to as a vesicular-arbuscular mycorrhiza (VAM). This name has since been dropped in favour of AM, since members of the Family Gigasporaceae do not form vesicles.

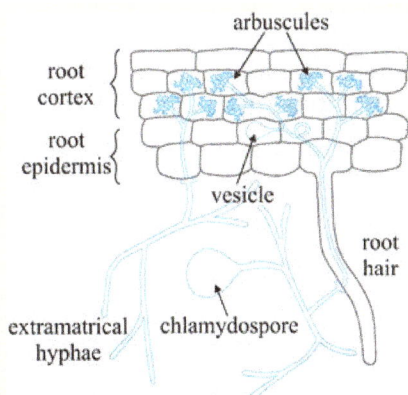

Diagrammatic representation of the main cellular features of the arbuscular endomycorrhiza. Hyphae from a germinating spore infect a root hair and can grow within the root between root cortical cells and also penetrate individual cells, forming arbuscules. These are finely branched clusters of hyphae, which are thought to be the major site of nutrient exchange between fungus and plant. For more information and illustration of mycorrhizas visit Mark Brundrett's. Figure and caption from Moore *et al.*, 2011 (URL).

There is a wide-ranging fungal mycelium within the host root, and AM fungi explore the soil or other substrata with an extensive extraradical mycelium. Externally to the host the fungal hyphae produce the very large spores (often called chlamydospores). Formation of the mycorrhizal association is an infection process. Spores germinate near a plant root and the germinating hyphae penetrate the root in response to root exudates. Hyphae grow through the root tissues and in the root cortex hyphal branches form appressoria that penetrate the plant cells. The host plasmalemma invaginates and proliferates around the fungal intrusion. Repeated dichotomous branching of the fungal 'hypha' produces the arbuscule inside the cortical cell. Arbuscules have a lifespan of 4-15 days, after which they break down, and the plant cell returns to normal.

A mature arbuscle of *Glomus mosseae* with numerous fine branch hyphae.

Physiology

Presymbiosis

The development of AM fungi prior to root colonization, known as presymbiosis, consists of three stages: spore germination, hyphal growth, host recognition and appressorium formation.

Spore Germination

Spores of the AM fungi are thick-walled multi-nucleate resting structures. The germination of the spore does not depend on the plant, as spores have been germinated under experimental conditions in the absence of plants both *in vitro* and in soil. However, the rate of germination can be increased by host root exudates. AM fungal spores germinate given suitable conditions of the soil matrix, temperature, carbon dioxide concentration, pH, and phosphorus concentration.

Hyphal Growth

The growth of AM hyphae through the soil is controlled by host root exudates known as strigolactones, and the soil phosphorus concentration. Low-phosphorus concentrations in the soil increase hyphal growth and branching as well as induce plant exudation of compounds that control hyphal branching intensity.

The branching of AM fungal hyphae grown in phosphorus media of 1 mM is significantly reduced, but the length of the germ tube and total hyphal growth were not affected. A concentration of 10 mM phosphorus inhibited both hyphal growth and branching. This phosphorus concentration occurs in natural soil conditions and could thus contribute to reduced mycorrhizal colonization.

Host Recognition

Root exudates from AMF host plants grown in a liquid medium with and without phosphorus have been shown to affect hyphal growth. Spores of Gigaspora margarita were grown in host plant exudates. Hyphae of fungi grown in the exudates from roots starved of phosphorus grew more and produced tertiary branches compared to those grown in exudates from plants given adequate phosphorus. When the growth-promoting root exudates were added in low concentration, the AM fungi produced scattered long branches. As the concentration of exudates was increased, the fungi produced more tightly clustered branches. At the highest-concentration arbuscules, the AMF structures of phosphorus exchange were formed.

This chemotaxic fungal response to the host plants exudates is thought to increase the efficacy of host root colonization in low-phosphorus soils. It is an adaptation for fungi to efficiently explore the soil in search of a suitable plant host.

Further evidence that arbuscular mycorrhizal fungi exhibit host-specific chemotaxis, that enable hyphal growth toward the roots of a potential host plant: Spores of Glomus mosseae were separated from the roots of a host plant, nonhost plants, and dead host plant by a membrane permeable only to hyphae. In the treatment with the host plant, the fungi crossed the membrane and always emerged within 800 μm of the root, but not in the treatments with nonhost plants and dead plants.

Molecular techniques have been used to understand the signaling pathways between arbuscular mycorrhizae and plant roots. In 2003 it was shown how the AM undergoes physiological changes in the presence of exudates from potential host plant roots, to colonize it. Host plant root exudates trigger and turn on AM fungal genes required for the respiration of spore carbon compounds. In experiments, transcription rate of 10 genes increased half-hour after exposure and at an even greater rate after 1 hour. after 4 hours exposure AM respond with morphological growth. Genes isolated from that time are involved in mitochondrial activity and enzyme production. The fungal respiration rate, measured by O_2 consumption rate, increased by 30% 3 hours after exposure to root exudates, indicating that host plant root exudates stimulate AMF spore mitochondrial activity. It may be part of a fungal regulatory mechanism that conserves spore energy for efficient growth and the hyphal branching upon receiving signals from a potential host plant.

Appressorium

When arbuscular mycorrhizal fungal hyphae encounter the root of a host plant, an appressorium or 'infection structure' forms on the root epidermis. From this structure hyphae can penetrate into the host's parenchyma cortex. AM need no chemical signals from the plant to form the appressoria. AM fungi could form appressoria on the cell walls of "ghost" cells in which the protoplast had been removed to eliminate signaling between the fungi and the plant host. However, the hyphae did not further penetrate the cells and grow in toward the root cortex, which indicates that signaling between symbionts is required for further growth once appressoria are formed.

Symbiosis

Once inside the parenchyma, the fungus forms highly branched structures for nutrient exchange with the plant called "arbuscules". These are the distinguishing structures of arbuscular mycorrhizal fungus. Arbuscules are the sites of exchange for phosphorus, carbon, water, and other nutrients. There are two forms: *Paris* type is characterized by the growth of hyphae from one cell to the next; and *Arum* type is characterized by the growth of hyphae in the space between plant cells. The choice between *Paris* type and *Arum* type is primarily determined by the host plant family, although some families or species are capable of either type.

The host plant exerts a control over the intercellular hyphal proliferation and arbuscule formation. There is a decondensation of the plant's chromatin, which indicates increased transcription of the plant's DNA in arbuscule-containing cells. Major modifications are required in the plant host cell to accommodate the arbuscules. The vacuoles shrink and other cellular organelles proliferate. The plant cell cytoskeleton is reorganized around the arbuscules.

There are two other types of hyphae that originate from the colonized host plant root. Once colonization has occurred, short-lived runner hyphae grow from the plant root into the soil. These are the hyphae that take up phosphorus and micronutrients, which are conferred to the plant. AM fungal hyphae have a high surface-to-volume ratio, making their absorptive ability greater than that of plant roots. AMF hyphae are also finer than roots and can enter into pores of the soil that are inaccessible to roots. The third type of AMF hyphae grows from the roots and colonizes other host plant roots. The three types of hyphae are morphologically distinct.

Nutrient Uptake and Exchange

AM fungi are obligate symbionts. They have limited saprobic ability and depend on the plant for their carbon nutrition. AM fungi take up the products of the plant host's photosynthesis as hexoses.

Carbon transfer from plant to fungi may occur through the arbuscules or intraradical hyphae. Secondary synthesis from the hexoses by AM occurs in the intraradical mycelium. Inside the mycelium, hexose is converted to trehalose and glycogen. Trehalose and glycogen are carbon storage forms that can be rapidly synthesized and degraded and may buffer the intracellular sugar concentrations. The intraradical hexose enters the oxidative pentose phosphate pathway, which produces pentose for nucleic acids.

Lipid biosynthesis also occurs in the intraradical mycelium. Lipids are then stored or exported to extraradical hyphae where they may be stored or metabolized. The breakdown of lipids into hexoses, known as gluconeogenesis, occurs in the extraradical mycelium. Approximately 25% of the carbon translocated from the plant to the fungi is stored in the extraradical hyphae. Up to 20% of the host plant's carbon may be transferred to the AM fungi. This represents the host plant's considerable carbon investment in mycorrhizal network and contribution to the below-ground organic carbon pool.

Increasing the plant's carbon supply to the AM fungi increases uptake and transfer of phosphorus from fungi to plant Likewise, phosphorus uptake and transfer is lowered when the photosynthate supplied to the fungi is decreased. Species of AMF differ in their abilities to supply the plant with phosphorus. In some cases, arbuscular mycorrhizae are poor symbionts, providing little phosphorus while taking relatively high amounts of carbon.

The main benefit of mycorrhizas to plants has been attributed to increased uptake of nutrients, especially phosphorus. This may be due to increased surface area in contact with soil, increased movement of nutrients into mycorrhizae, a modified root environment, and increased storage. Mycorrhizas can be much more efficient than plant roots at taking up phosphorus. Phosphorus travels to the root or via diffusion and hyphae reduce the distance required for diffusion, thus increasing uptake. The rate of phosphorus flowing into mycorrhizae can be up to six times that of the root hairs. In some cases, the role of phosphorus uptake can be completely taken over by the mycorrhizal network, and all of the plant's phosphorus may be of hyphal origin. Less is known about the role of nitrogen nutrition in the arbuscular mycorrhizal system and its impact on the symbiosis and community. While significant advances have been made in elucidating the mechanisms of this complex interaction, much investigation remains to be done.

Mycorrhizal activity increases the phosphorus concentration available in the rhizosphere. Mycorrhizae lower the root zone pH by selective uptake of NH_4^+ (ammonium-ions) and by releasing H^+ ions. Decreased soil pH increases the solubility of phosphorus precipitates. The hyphal NH_4^+ uptake also increases the nitrogen flow to the plant as the soil's inner surfaces absorb ammonium and distribute it by diffusion.

Arbuscular mycorrhizal (AM) fungi interact with a wide variety of organisms during all stages of their life. Some of these interactions such as grazing of the external mycelium are detrimental, while others including interactions with plant growth promoting rhizobacteria (PG PR) promote mycorrhizal functioning. Following mycorrhizal colonisation the functions of the root become modified, with consequences for the rhizosphere community which is extended into the mycorrhizosphere due to the presence of the AM external mycelium. However, we still know relatively little of the ecology of AM fungi and, in particular, the mycelium network under natural conditions. This area merits attention in the future with emphasis on the fungal partner in the association rather than the plant which has been the focus in the past.

Taxonomy and Species Recognition

The arbuscular mycorrhizal fungi belong to the Glomeromycota. They were formerly placed in the Zygomycota (Thaxter, 1922; Gerdemann & Trappe, 1974), but molecular analyses suggest that they should have their own phylum. The affinity to Zygomycota was based on the interpretation of the spores as azygospores, which are zygospores with only one gametangium. Later, the spores were considered to be a sporangium in which the spores do not develop, or a merosporangium with only one spore. The homology of the spores to other known fungal structures has never been established, and recent studies suggest that the Zygomycota are polyphyletic. At present the Glomeromycota has no obvious affinity to other major phylogenetic groups in the kingdom Fungi (James *et al.*, 2006). The interpretation of homologies between Glomeromycota spores and other fungal structures is further complicated by the fact that Glomeromycota spores may, among themselves, represent different morphological and functional structures. In some *Glomus* species such as *G. intraradices*, the structure of the spore wall is simple, whereas the spores in *Scutellospora* spp. can have several inner membranous walls and develop complex germination shields. The finding of dimorphic species with both *Glomus* and *Acaulospora* morphs further suggests that all spores in the Glomeromycota may not represent homologous structures (Morton & Redecker, 2001).

The first classification systems were not always based on the Botanical Code, but used descriptive names such as yellow vacuolated (YV) for *Glomus mosseae*, honey-coloured sessile for *Acaulospora laevis*, etc. (Mosse & Bowen, 1968); or E1, E2, E3, etc. (Gilmore, 1968). These descriptive names were often mistaken for isolate or strain numbers. For example, the famous E3 isolate from Rothamsted Experimental Station only resembled Gilmore's E3 from California, but was actually isolated in the UK. The first taxonomic monograph was published in 1974 by Gerdemann & Trappe (1974), but no attempts to monograph the Glomeromycota have been made since that time. As an alternative, N. Schenck and Y. Perez at the University of Florida collected the species descriptions in a manual that was used at the taxonomic workshops in the 1980s and 1990s. A major breakthrough in species identification has been the establishment of culture collections and reference strains: Banque Européenne des Glomales (BEG) in Europe and the International Culture Collection of Vesicular (Arbuscular) Mycorrhizal Fungi (INVAM) in North America. The descriptions found on the INVAM, provided by Joe Morton, have served as a reference for species designation by researchers worldwide. The descriptions are based on INVAM cultures, and do not attempt to cover the variation in morphological characters. Although there are problems with correct identification of species in various culture collections, these collections have greatly facilitated comparisons among experiments where such *bona fide* cultures are used.

The taxonomic system was based on interpretation of morphological features of the resting spores, and later studies have included characters related to the ontogeny of the spores (Franke & Morton, 1994). Interpretation of these characters has laid the ground for several controversies among taxonomists, a phenomenon not specific to mycorrhizal research. The main discussion has focused on the spore wall characters, where the relevance of the number and position of walls and wall layers has been a key issue. A so-called murograph was included in the description of some species, and new species were described containing up to seven walls. Several users of the classification found it very hard to identify species based on the descriptions (Koske, 1986), and some claimed that the number of walls depended more on the age of the spore and how hard you pressed the cover slip, than on the actual taxonomic identity (J. C. Dodd, personal communication). The discussion on whether arbuscular mycorrhizal fungi had several walls (*sensu* Walker, 1983) or one wall with several layers (Berch, 1987) may have been regarded as a purely academic debate, but the discussion reflected the question of whether or not the diversity had a common origin, or if the wall characters were independent characters. Later studies of wall ontogeny of *Scutellospora pellucida* showed that walls or wall layers cannot be seen as independent characters, as the inner walls differentiate during germination of the spore (Franke & Morton, 1994). On several occasions Morton has made a plea for a hierarchical interpretation of taxonomic characters (Morton *et al.*, 1992; Morton 1995), but species are still described without such interpretation of the characters.

The number of recognizable species in the Glomeromycota is not apparent. Despite the worldwide distribution of the fungi, relatively few species have been described. Some authors recognize *c.* 200 species, but some additional descriptions exist and some of the descriptions could be synonyms. Recently, A. Schüßler in Munich listed all described species in Glomeromycota. There are 104 described species in the genus *Glomus* alone, but interpretation of some of these descriptions can be problematic. For some species the herbarium specimens are unavailable and it is not possible to verify the descriptions; in other cases the material is in a very bad condition and the taxonomic features are no longer recognizable. This is often combined with old, rather vague descriptions of the material, leaving interpretation of key characters open for discussion.

Species Concepts and Species Recognition

The number of species depends to a great extent on the concept used to recognize the species. Most biologists agree on an evolutionary species concept based on Darwinian evolution that defines species as: 'a single lineage of ancestor-descendent populations which maintains its identity from other such lineages and which has its own evolutionary tendencies and historical fate' (Wiley, 1978). However, the evolutionary species concept is a theoretical concept, and cannot be used to recognize or diagnose species. Instead, morphological, biological and phylogenetic species concepts can be regarded as operational concepts, as they all intend to recognize evolutionary species (Mayden, 1997). Based on this, Taylor *et al.* (2000) used the term 'species concept' for the theoretical concept and the term 'species recognition' for the operational concepts.

Taylor *et al.* (2000) discussed various species recognition concepts in fungi and proposed the Genealogical Concordance Phylogenetic Species Recognition (GCPSR). The basic idea of GCPSR is that in a phylogenetic tree, concordant branches represent species, whereas incongruity caused by recombination is found within species. They state that a drawback of the concept is that it will not be useful for strictly clonal organisms, but claim that such organisms may not exist, and that the studies of primarily clonal fungi have shown that recombination takes place. They conclude that if true clonal fungi exist, they are very rare, but they do not consider the Glomeromycota. Recombination has not yet been proved in the Glomeromycota, and most evidence suggests that the fungi have evolved asexually. The lack of incongruity to delimit species means that GCPRS cannot be used to recognize arbuscular mycorrhizal fungal species.

Presently, there is no formalized operational species concept in Glomeromycota. Species are described based on morphology and information from sequences of ribosomal RNA genes (rDNA). The access to sequence information has not changed the criteria for species recognition in the Glomeromycota, and studies based on rDNA have often confirmed the morphologically defined species. Some new descriptions of Glomeromycota taxa include sequence information, which is particularly valuable for delimiting taxa with few available morphological characters. The molecular data have substantially changed the systematic, and several new genera and families have been erected.

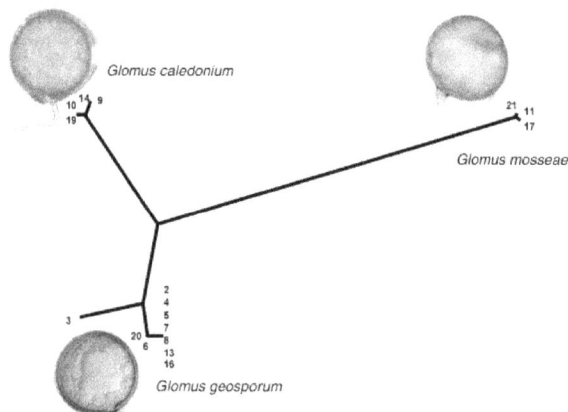

Figure: Unrooted parsimonious tree showing three closely related *Glomus*species isolated from the same field in Denmark. The analysis is based on sequences of LSU rDNA and introns in the protein-coding genes *TOR1* and *FOX1* amplified from single spores

I. Communities of Arbuscular Mycorrhizal Fungi

Previously, community studies of arbuscular mycorrhizal fungi relied on the presence of resting spores in the soil. These spore surveys are often criticized, as it cannot be determined if species are truly absent or they are present but not sporulating. A comparison of communities is thus not possible if the communities are dominated by nonsporulating fungi. For example, Sparling & Tinker (1975) observed fewer spores in permanent vegetation and suggested that sporulation is related to disturbed systems. Quantitative studies of arbuscular mycorrhizal fungal communities based on the presence of spore numbers are also complicated as some species produce few spores on the mycelium, whereas cluster-forming species such as *G. intraradices, G. versiforme* or *G. fasciculatum*produce hundreds of spores on the same hypha. Although the spore surveys may not provide sufficient information on communities of arbuscular mycorrhizal fungi, they have revealed interesting patterns (Walker *et al.*, 1982; Bever *et al.*, 2001). A problem with the previous spore-based community studies is that they are sometimes difficult to interpret and relate to more recent studies, as the species identification is uncertain and may have changed over time. In several studies from the 1970s and 1980s, *Glomus fasciculatum* appears to be a very common species, but after the species was redescribed (Walker & Koske, 1987) it was shown to be uncommon; reports of *G. fasciculatum* in older surveys and experiments must have been of other species. Application of molecular techniques has made such comparisons more reliable as sequences are deposited in GenBank and can be compared with other known species.

1. Species known only from Environmental Samples

The molecular approach to community ecology of arbuscular mycorrhizal fungi has revealed a large unknown diversity, and has shown that known species sometimes can account for only a minor proportion of the diversity. In a community study from a perennial grassland in Denmark, four out of 11 phylogenetic clusters in *Glomus* belonged to species where only sequence information is available (Rosendahl & Stukenbrock, 2004), and only one sequence type of 10 was identified from another site in a study from Scotland (Gollotte *et al.*, 2004). These studies are based on large subunit (LSU) rDNA, whereas most studies are based on small subunit (SSU) rDNA. The resolution of these genes is different, and a direct comparison of phylogenetically defined taxa is not possible. Closely related taxa such as *G. caledonium* and *G. geosporum* have only 2- or 3-bp substitutions in SSU, but can be easily separated by their morphology and their LSU rDNA sequence.

In phylogenetic lineages with no known morphologically defined species, it is not possible to make this evaluation. Often lineages are defined as taxa, but the taxa defined solely from trees based on single genes should be referred to as phylogenetic clusters rather than phylogenetic species, as the later implies species defined based on gene concordances (Taylor *et al.*, 2000). Divergence in one locus may give a history only of the locus, not of the species, as different genes or gene regions may be subjected to different mutation rate, selection pressure, etc. To detect cryptic species, phylogenies inferred from more than one genomic region (multilocus DNA sequence phylogenies) must be used. Such phylogenies should detect speciation before the whole genome has diverged (Kohn, 2005).

The gene tree approach to taxon recognition is sensitive to sampling bias. Often only sequences from a particular location are included in the phylogenetic analysis. If some taxa are poorly represented in an area, or intermediate phylogenetic lineages are missing, artificial high similarity between distant taxa may occur. In a community study from Denmark (Rosendahl & Stukenbrock, 2004), where samples were taken as individual root fragments, some groups were clearly better represented than others. In the phylogenetic lineage with *G. intraradices* (often referred to as *Glomus* Ab), several sequences were found only once. If the level of sequence divergence observed between the morphologically defined species *G. caledonium*, *G. mosseae* and *G. geosporum* is used to delimit species, at least 15 phylogenetic taxa can be defined in this study. All sequences fall within five main lineages, but interestingly two of the main lineages (I and II) have no known morphologically defined species.

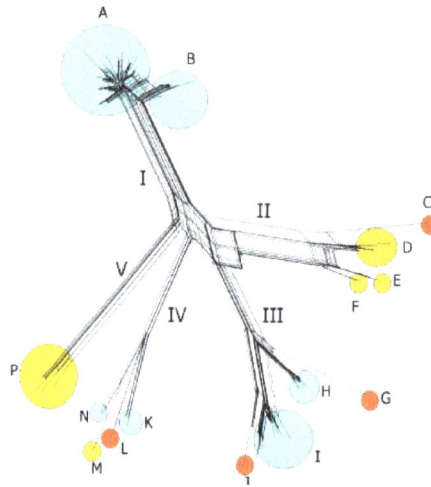

Figure: NeighborNet split network based on least-square distances of 300 LSU rDNA sequences obtained from single-root fragment in a 0.1-km² coastal grassland in Denmark (Rosendahl & Stukenbrock, 2004). Yellow circles are sequence types of known species: D, *G. mosseae*; E, *G. caledonium*; F, *G. geosporum*; M, *G. intraradices*, P, *G. microaggregatum*. Red circles are unique sequences (singletons); blue circles are clusters with no known related species.

2. Species Abundance and Distribution

From the community studies, it is obvious that communities of arbuscular mycorrhizal fungi follow the same distribution pattern as many other organisms, with few very common species, and a number of more and more rare species. Which fungi are common and which are rare depends on the habitat. We still know little about the autecology of arbuscular mycorrhizal fungi, but from several observations it can be concluded that fungi such as *G. mosseae*,*G. caledonium*, *G. geosporum* and *Scutello-*

spora pellucida are common in arable soils, whereas the patterns in woodlands and uncultivated grasslands are less apparent. This is mainly because very few such systems have been studied. In their overview of studies of global arbuscular mycorrhizal fungal communities,Öpik *et al.* (2006) conclude that some fungi could be specific to forests. However, this comparison was based on very few studies, and there was an obvious correlation between the number of studies at different sites and the number of phylogenetic groups found.

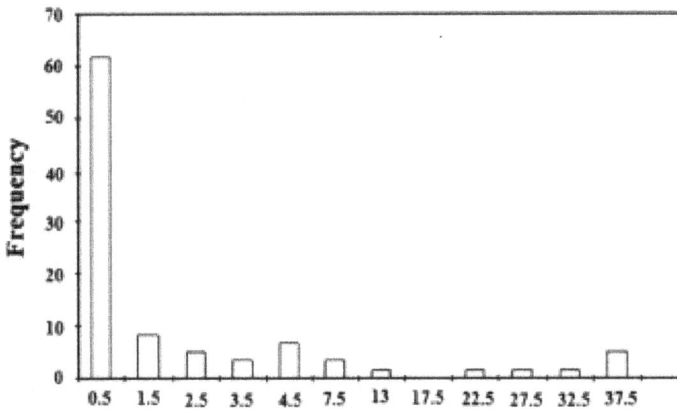

Figure: Distribution of sequence types in Figure,
showing that the community is dominated by a single type

The development of molecular tools to characterize microbial communities has resulted in a considerable increase in the number of community studies of arbuscular mycorrhizal fungi. The fungi have been identified using either sequencing (Van Tuinen *et al.*, 1998) or denaturing gradient gel electrophoresis (DGGE) and terminal-restriction fragment length polymorphism (T-RFLP) approaches (de Souza *et al.*, 2004; Lekberg *et al.*, 2007) following the isolation and PCR of fungal DNA. In studies where Glomeromycota DNA from roots has been sequenced, the ribosomal genes have been targeted: either the SSU, first used by the group at York University (Helgason *et al.*, 1998, 2002); or the LSU, used in Dijon (Van Tuinen *et al.*, 1998) and Copenhagen (Kjøller & Rosendahl, 2000). In many studies, only one pair of primers (NS31 and AM1) for SSU was used to cover the diversity in Glomeromycota (Helgason *et al.*, 1998). This primer pair will amplify most but not all groups (Redecker *et al.*, 2003), and several nonglomalean sequences may be amplified (Douhan *et al.*, 2005), making subsequent use of DGGE and T-RFLP problematic. The LSU provides a better resolution, but several primers are necessary for amplifying all genera of Glomeromycota. Another important disadvantage of targeting the ribosomal genes is that copies of the multi-copy gene may vary within a single spore, making subsequent cloning necessary (Scheublin *et al.*, 2004). If the amplified PCR products are cloned, the quantitative aspect is lost, which can be important when host preferences or responses to environmental gradients are studied. To retain the quantitative aspect, LSU rDNA was amplified from 1-cm root fragments and sequenced directly, and the frequency of sequence types was quantified as the number of colonized root fragments (Rosendahl & Stukenbrock, 2004; Stukenbrock & Rosendahl, 2005b). Although both approaches have drawbacks, the results have been remarkably similar, showing a limited diversity of arbuscular mycorrhizal fungi in most vegetation systems.

The species richness in arbuscular mycorrhizal fungi communities is not easy to estimate because of problems in identifying nonsporulating species. A study of spore diversity in a field in North Carolina revealed 37 species recognized as spores either sampled directly or trapped on plants (Bever *et al.*, 2001). This relatively high species richness probably reflects the sampling intensity, rather than characteristics of the area. The authors made no attempt to extrapolate from the study, but noted that the richness in the field was similar to what has been reported from continents. Öpik *et al.* (2006) compared most community studies and found that the average number of species (taxa) ranged from 1 to 29. However, the criterion for species recognition is crucial for interpreting and comparing community studies as well as the different genetic markers used to identify species.

II. Populations of Arbuscular Mycorrhizal Fungi

The concept of a population is used widely in ecology, population genetics and evolutionary biology. Several definitions can be found, but the central theme is that a population is used to describe a group of individuals of a certain species. Two main paradigms exist: the ecological and the evolutionary. The ecological paradigm identifies a population as: 'a group of organisms of the same species occupying a particular space at a particular time' (Krebs, 1994). Although this definition may sound precise, the problem is that different researchers may look at different space and time scales, and thus not draw the same conclusions. To overcome this problem, the ecological population concept can be amended by adding: 'and have the opportunity to interact with each other' (Waples & Gaggiotti, 2006). The evolutionary paradigm is centred on the definition by Dobzhansky (1937): 'a community of individuals of a sexually reproducing species within which mating takes place'. This definition emphasizes the reproductive interaction between the individuals, and has some obvious limitations when considering fungal populations. The broader definition by Waples & Gaggiotti (2006), which identifies a population as 'a group of individuals of the same species living in close enough proximity that any member of the group can potentially mate with any other member', is more appropriate to mycology. The main idea in the evolutionary paradigm is that populations can be characterized by their allele diversity. Differences in allele frequencies reflect subdivisions between populations. Gene flow is the main introgression of new alleles into a population and the main mechanism to reduce genetic differentiation between populations. However, in strictly asexual organisms the genes are linked and can only move as a block. This type of gene flow is better termed 'genotype flow'. In arbuscular mycorrhizal fungi without recombination, it makes no sense to talk about gene flow, as gene flow considers single alleles. With complete linkage of alleles, the individuals will migrate and will keep their genetic integrity.

1. Global Migration or Endemism

Populations of microbial eukaryotes are believed to cover continents because of the effective dispersal of airborne spores. This is often referred to as the 'ubiquitous dispersal hypothesis' (Finlay, 2002; Fenchel & Finlay, 2004), and is based on the famous citation 'everything is everywhere, but the environment selects' (Baas Becking, 1934). The hypothesis suggests that fungi have large effective population size, and that genetic drift plays a minor role in evolution of the fungi. The hypothesis is controversial and has been challenged by population genetic studies of cosmopolitan fungi, demonstrating that what may look as cosmopolitan is actually hidden endemism (Taylor et al., 2006). Several species in the Glomeromycota are known to occur on different continents supporting the ubiquitous dispersal hypothesis. However, genetic differentiation between geographical isolates has not been studied, and it is not known if the morphologically recognized species hide endemic cryptic species. Alternatively, the apparent global distribution of some arbuscular mycorrhizal fungi species could be a recent phenomenon caused by human activity related to agriculture. Several studies have demonstrated global migration among fungal pathogen populations (Goodwin et al., 1994; Zhan et al., 2003).

Recombination and gene flow are important factors in shaping the genetic structure of populations. Asexual fungi evolve without the homogenizing effect of recombination and gene flow. The individual genotypes within a population can evolve differently by the accumulation of mutations in loci that are not under selection. In time, this genetic differentiation of clonal individuals can lead to a phenotypic or functional differentiation. Dispersal becomes important in fungi where individuals that differ from the most frequent genotype have a selective advantage and can establish or even become dominating in an area. Among pathogens, such asexual dispersal may initiate epidemics and even pandemics. The human pathogen *Penicillium manefii* is an example of an asexual fungus in which, despite widespread aerial dispersal, isolates of the species showed extensive spatial genetic structure at local and country-wide scales (Fisher et al., 2005). The authors suggested that this endemism is a consequence of the lack of sexual reproduction that has led to evolution of niche-adapted genotypes despite the extensive aerial dispersal of the pathogen. Their results are interesting when considering arbuscular mycorrhizal fungi. If asexuality has led to the evolution of niche-adapted genotypes, it is reducing the potential of the fungus to diversify. In a heterogeneous environment, variance in fitness between clonal lineages is expected to lead to populations that are adapted to local conditions (Goddard et al., 2005). In this case, limited dispersal becomes an inevitable consequence of asexuality, as invading clones are outcompeted by better-adapted local competitors because of the decreased ability of invaders to diversify within novel environments (Buckling et al., 2003).

2. Ancient Origin without Recombination

Despite an early observation of zygospore formation in *Gigaspora* (Tommerup & Sivasithamparam, 1990), arbuscular mycorrhizal fungi are generally believed to be asexual. Recombination can be shown by analyses of linkage of multiple alleles (index of association) (Maynard Smith *et al.*, 1993). This has been used to demonstrate recombination in *Coccidium immitis*, where no sexual state has been observed (Burt *et al.*, 1996). In arbuscular mycorrhizal fungi, analysis has not so far revealed signs of recombination in data obtained from pot cultures or field-collected spores (Rosendahl & Taylor, 1997; Stukenbrock & Rosendahl, 2005a). A single study found recombination in a data set based on inter simple sequence repeat (ISSR) markers obtained from field-collected spores (Vandenkoornhuyse *et al.*, 2001). However, dominant markers such as ISSR are problematic for estimating linkage, as the homology between the alleles is difficult to verify. If the alleles are not homologous, analyses such as index of association (Maynard Smith *et al.*, 1993) may suggest recombination erroneously. A study by Gandolfi *et al.* (2003) used haplotype networks to demonstrate that DNA sequences from the same *Glomus* species may show signs of recombination. Unfortunately, the data were based on sequences from GenBank, and some of the data sets contained sequences from more than one species. Some of these species are only distantly related, and the reticulate structure of the haplotype networks is more easily explained by a saturation of mutations rather than recombination.

Linkage of alleles indicates lack of recombination, but will not prove that the organisms are ancient asexuals. Normark *et al.* (2003) hypothesized that a clonal population structure would appear after few asexual generations if the effective population size (N_e) is low. This was based on the assumption that the coalescent time since the most common recent ancestor is approximately $2N_e$ generations (Birky, 1996). The coalescence model assumes a neutral model, and the actual coalescent time would be less if the loci are linked. We know little about the population size and generation time of arbuscular mycorrhizal fungi, but if the population size is low, the recombining structure of populations should be lost in few thousand generations (Normark *et al.*, 2003).

A genomic signature that would advocate for an ancient asexual evolution would be strong divergence between alleles in the same locus, as has been shown for diploid bdelloid rotifers (Mark Welch & Meselson, 2000). However this would only be the case with diploid organisms, as the pattern of divergence is equally consistent with the organism being sexual and haploid. Strong divergence has been observed in genes that may have a common origin and may suggest an ancient asexual evolution of arbuscular mycorrhizal fungi (Kuhn *et al.*, 2001; Pawlowska & Taylor, 2004).

Evidence against an ancient asexual evolution would be the existence of genes involved in sexual functions. Preliminary studies have found possible candidate genes, and this may change our view of arbuscular mycorrhizal fungi as ancient asexuals (Colard *et al.*, 2007). Further evidence of recent meiotic events would be the presence of selfish genetic elements or retrotransposons, which are believed to disappear in the absence of meiosis (Hickey, 1982; Arkhipova & Meselson, 2000). Retrotransposons have been detected in Glomeromycota genomes, but contained stop codons and were not expressed in either hyphae or germinating spores (Gollotte *et al.*, 2006).

3. Genetic structure of arbuscular mycorrhizal fungi populations

The genetic structure of a population refers to the distribution of genetic variation, and can be studied by hierarchical sampling and analysis of genetic variation within and between fields, plots, subplots, etc. Nei's (1987) analysis of genetic variation and analysis of molecular variance (AMOVA) (Excoffier *et al.*, 1992) are two methods for estimating population differentiation directly from molecular data and for testing hypotheses regarding factors that may cause such differentiation.

AFLP data analysed by AMOVA (Koch *et al.*, 2004) and multilocus genotyping of arbuscular mycorrhizal spore populations in agricultural systems, analysed using Nei's approach (Stukenbrock & Rosendahl, 2005a), showed a patchy distribution of genotypes within populations of common mycorrhizal species. A patchy distribution of arbuscular mycorrhizal populations will arise when genotypes in the field are structured as distinct mycelia that are able to establish and maintain their genetic integrity. In an agricultural system with annual crops, the turnover of root systems requires germination of new mycorrhizal spores every year. In a model that will explain the observed pattern, the spores germinate in the spring and hyphae of the same genotype fuse into a common mycelial network. This will result in genetically uniform hyphal networks producing new spores of the same genotype. Vegetative incompatibility between *Glomus* isolates from geographically different areas has been reported by Giovannetti *et al.* (2003), also demonstrating a high frequency of self-fusion between hyphae of the same isolates. If self-recognition in arbuscular mycorrhizal fungi ensures hyphal fusion only between hyphae of the same genotype, this mechanism may be regarded as a genetic bottleneck, excluding rare or mutated genotypes from the mycelial networks. It is, however, important to note that the genetic background for incompatibility in arbuscular mycorrhizal fungi is not known, and could be different from what has been studied in Ascomycota (Glass *et al.*, 2004).

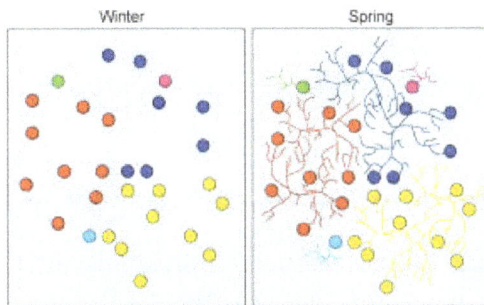

Hypothesis explaining how mycelial networks are formed by arbuscular mycorrhizal fungi in disturbed systems such as agroecosystems. In winter, resting spores are distributed in the soil. Different genotypes are represented by different colours. The population consists of three common genotypes (blue, red, yellow) and three rare genotypes (green, pink, turquoise). In spring the spores have germinated and mycelia with the same genotypes anastomose and form hyphal networks. Rare and unique genotypes will not anastomose and the dominant genotypes will be favoured.

III. Individuals, Genets or Clones

We know very little about what constitutes the individual mycelium of an arbuscular mycorrhizal fungus. Several authors have discussed individuality in fungi, but no conceptual agreement exists. This is partly because individuals can be defined in both genetic and physiological terms. The mycelium that grows from a spore (sexual or asexual) or a hyphal fragment will, in physical or physiological terms, constitute an individual. However, the genetic or evolutionary unit will comprise genetically identical fragments or spores derived from mating between two nuclei. Such units can be defined as genets. In plants, genets are defined as all vegetative derivatives of a seed, including vegetative propagated plants from stolons or rhizomes. In mycology, the same definition can be used for many basidiomycetes with dikaryotic or diploid mycelia. In Basidiomycota, fusion between two haploid gametes will form a new dikaryotic mycelium. Even if the fungus may have subsequent asexual propagation, the genets are similar to plant genets in both functional and genetic aspects. This definition is not directly applicable for fungi with haploid mycelia such as the Ascomycota, Zygomycota and Glomeromycota (Hijri & Sanders, 2004). In Ascomycota, fusion of gametes or gametangia will result in dikaryotic ascogenic hyphae that will form the asci where meiosis takes place. This structure is often part of a fruiting body (ascoma), which is neither a functional nor an evolutionary unit. Anderson & Kohn (1995) proposed that genets of haploid fungi (e.g. Ascomycota) are defined as all vegetative derivatives of the fusion of two genetically unlike gametic nuclei and the following meiosis. For fungi with asexual reproduction, genets can be composed of several functional units, but they will all have the same origin. If the origin is not known, the term 'clone' is used to cover organisms with the same multilocus genotype. The term clone often refers to genetically identical individuals, but clones of ancient origin might accumulate mutations during mitotic cycles. Such nearly identical clones can be defined as belonging to the same clonal lineage. Identification of such lineages can be difficult if the organisms are ancient asexuals and/or distributed over wide areas, as the organisms may develop new phenotypes and genetic differences that could be interpreted as representing more than one genet.

1. Recognizing Genets and Clones of Arbuscular Mycorrhizal Fungi

Very few authors have addressed the issue of individuals and genets of arbuscular mycorrhizal fungi. Again, it is important to distinguish between functional or physical coherent mycelia and genetic evolutionary units. Studies have been made of the extension and growth rate of individual hyphae (Mosse et al., 1982; Warner & Mosse, 1983) suggesting that arbuscular mycorrhizal fungi could be capable of forming extensive coherent mycelia. However, it is not known if the mycelia are coherent or fragmented into smaller ramets. Anastomosis in arbuscular mycorrhizal mycelia can be observed in roots as H-connection, and between hyphae of germinating spores (Mosse, 1959). These anastomoses may result in a reticulate growth of the mycelium, and may also allow possible nuclear migrations between individual mycelia. The experimental system set up by Manuela Giovannetti and coworkers in Pisa has made it possible to observe hyphal anastomosis *in vivo*, and revealed that the germinating hyphae of *G. mosseae* and other arbuscular mycorrhizal fungi anastomose shortly after germination. In a later study they demonstrated that mycelia colonizing different host plants form anastomoses, which makes it possible to have an infinite hyphal network that connects different plant species. Nuclear migration through the anastomoses was also observed, but a study using isolates from different geographical locations showed that anastomoses occurred only within the same isolate of *G. mosseae*. The studies were conducted under laboratory conditions and therefore cannot be translated to field conditions, but the potential of forming hyphal networks has led several authors to discuss their role in nutrient acquisition of the plant.

If arbuscular mycorrhizal fungi grow as fragmented mycelia all belonging to the same clone, such mycelia may not differ phenotypically. This raises the question of functional redundancy – a subject that has been discussed by several authors. Fitter (2005), in his presidential address, discusses functional redundancy and comes to the conclusion that the concept of functional redundancy may itself be redundant, as the question can be reduced to: what determines the ability of numerous species with similar ecological function to coexist? The question is highly relevant for Glomeromycota, as the coexistence of clonal lineages of the same phylogenetic type or species is a fact within fields (Stukenbrock & Rosendahl, 2005a). Functional differences between arbuscular mycorrhizal fungi may exist, but so far it has not been possible to assign specific functions to species or isolates. A large variation was seen in phosphorus uptake and the amount of external mycelium within isolates of both *G. mosseae* and *G. caledonium*, but isolates within each species were strikingly similar in their length-specific hyphal phosphorus uptake. (Munkvold *et al.*, 2004). Furthermore, isolates of *G. intraradices* from one field differed in their effect on plant growth (Koch *et al.*, 2006). This indicates that the functional diversity could be structured in the field, but in order to study this a more detailed sampling of species and isolates from various spatial scales is required.

2. Homokaryons, Heterokaryons or Polyploids

Some authors have claimed that the arbuscular mycorrhizal fungi represent multigenomic organisms (Kuhn *et al.*, 2001; Sanders, 2002), and it has been hypothesized that this heterokaryotic structure has arisen by hyphal anastomosis between genetically different mycelia and by accumulation of mutations (Bever & Wang, 2005). The issue is controversial, and raises several questions regarding the ecology and evolution of such multigenomic organisms. Studies of arbuscular mycorrhizal fungi *in situ* have not confirmed the multigenomic status of the fungi. Instead, the fungi seem to form discrete and genotypic uniform mycelia.

The existence of high genetic variability within spores cannot be questioned, and multiple variants of ribosomal genes are known to occur within single spores (Sanders *et al.*, 1995; Corradi *et al.*, 2007). However, whether the observed polymorphism is structured between genetically different nuclei (Kuhn *et al.*, 2001), between different sets of chromosomes (Pawlowska & Taylor, 2004), or within each nucleus as duplicated genes (Hosny *et al.*, 1999; Rosendahl & Stukenbrock, 2004) is a matter of debate. In sexual organisms, tandemly repeated copies of rDNA genes are homogenized by concerted evolution, a process where unequal crossing-over and gene conversion will homogenize sequences during meiosis. The lack of meiosis excludes concerted evolution and leads to different gene variants in genomes, as observed in other asexual organisms (Gandolfi *et al.*, 2001).

The hypothesis of multigenomic mycorrhizal mycelia raises several questions regarding the biology and evolution of such organisms. Hyphal fusion and heterokaryon formation is a known phenomenon in filamentous Ascomycota (Glass *et al.*, 2004). Although there are apparent benefits associated with heterokaryon formation, heterokaryosis by hyphal fusion is believed to be virtually excluded in nature by genetic differences at heterokaryon incompatibility loci. This nonself recognition mechanism prevents the establishment of a compatible heterokaryon between genetically different individuals. The patchy distribution of fungal genotypes in nature may reflect this nonself recognition mechanism, which could ensure genetic integrity of the mycelia.

The possible existence of heterokaryosis in the mycelia of arbuscular mycorrhizal fungi raises another important question. How would gene regulation be coordinated between the populations of genetically different nuclei in one continuous mycelium? In studies regarding patterns of gene expression, heterokaryosis has not been reported. It is striking that when the gene diversity is studied from markers developed from cDNA, no sign of heterokaryosis is seen, whereas markers developed from degenerated primers result in several gene variants within each spore. Future research on arbuscular mycorrhizal fungi should focus not only on the genetic background of hyphal anastomosis, but also on variation in cDNA libraries within and between populations of different mycorrhizal genotypes.

IV. Speciation in Glomeromycota

Species are regarded as a fundamental unit in biology, yet the true existence of such units can be questioned. Speciation has been reviewed by several authors, but few reviews focus on speciation in fungi (Kohn, 2005). As stated earlier, reproductive isolation is believed to be an important factor for speciation in sexual species. Within the species, interbreeding will unify the individuals, and the restricted gene flow between species will allow the species to diverge. In asexual organisms, this unifying effect of interbreeding does not exist, and it has been postulated that asexual species do not exist as the individual is the only evolutionary unit (Coyne & Orr, 1998). Whether speciation occurs in asexual organisms is still unclear, but if reproductive isolation cannot explain the existence of species, then adaptation to specific niches must be a crucial factor in speciation (Coyne & Orr, 1998). This suggests that speciation can occur sympatrically in different niches, and that asexual species can be revealed as phylogenetically distinct clades. Most asexual organisms are probably of recent origin, and it is difficult to study the processes of speciation. Asexual ascomycetes such as *Penicillium* and *Aspergillus* have closely related sexual relatives, and their origins are probably due to loss of a sexual cycle (LoBuglio *et al.*, 1993). Other results indicate that asexuality in the rice blast pathogen *Magnaporthe oryzae*could be the result of loss of a mating type during adaptation to rice, and thereby also a recent event related to the onset of agriculture in Asia (Couch *et al.*, 2005). An exception is the previously mentioned asexual bdelloid rotifer that may be more than 100 million yr old. Interestingly, recent results show that these organisms diversify into entities similar to sexual organisms, and raise the question as to whether sex is necessary for speciation (Fontaneto *et al.*, 2007). The Glomeromycota have no known closely related sexual species and, as discussed earlier, their asexuality could also be of ancient origin. Similar to the bdelloid rotifers, morphological species do exist, and the species are supported by phylogenetic analyses of molecular data. The closely related species *G. mosseae*, *G. geosporum* and *G. caledonium* clearly represent entities with a morphological and genetic integrity. The age of the divergence of these species is not known, but it could be between 10 and 100 million yr. If speciation occurs without recombination and with complete linkage of genes, selective sweeps must have occurred at various times. Such selective sweeps may make intermediate lineages extinct, leaving lineages that are genetically distant. If the linkage of genes has resulted in the evolution of genetically distant individuals that can be regarded as physiological species, the phylogenetic and morphological species that we recognize today may represent a much higher level in the genealogical hierarchy compared with other fungi.

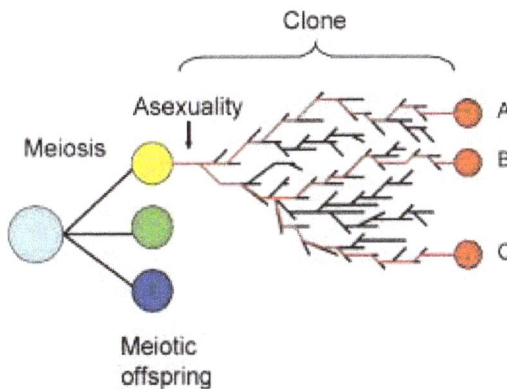

Figure: A long asexual evolution may result in very different genotypes belonging to the same clone. Complete linkage of loci will result in selective sweeps that will make intermediate genotypes extinct, leaving distantly related genotypes. A–C, recent genotypes that are recognized as isolates, species or even genera depending on the time scale of the ancient asexual evolution.

The genetic differentiation in *Glomus* spp. populations observed within fields can be the result of niche differentiation in a heterogeneous soil environment. Both experimental and retrospective studies support this hypothesis. An arbuscular mycorrhizal community assembly was shown experimentally to depend on the phylogenetic relatedness of species, suggesting that reduced competition between distinct evolutionary lineages promotes their coexistence (Maherali & Klironomos, 2007). This experimental study was conducted with phylogenetically distant arbuscular mycorrhizal fungi, but a similar pattern was observed in an analysis of 400 *Glomus sensu stricto* sequences obtained from perennial grassland in Denmark (Rosendahl & Stukenbrock, 2004), where the sequences were organized in a distinct pattern with five major lineages above species level. However, the study also revealed that there were several sequence types coexisting within these clades. Similarly, a study from an agricultural field showed that several genotypes of three *Glomus* species were able to coexist (Stukenbrock & Rosendahl, 2005a). Whether these genotypes compete is not known, but other studies have shown that isolates within the same species can be phenotypically different, which may allow them to coexist in the same habitat (Koch *et al.*, 2004; Munkvold *et al.*, 2004). Although it may be tempting to suggest that the haplotypes have evolved as a result of niche differentiation, the number of mutation suggests that the diversification is ancient, at least exceeding the age of Denmark (approx. 10 000 yr). The diversification must therefore have occurred in a different environment, and the haplotypes colonized the field later.

In conclusion Arbuscular mycorrhizal fungi are often referred to as mysterious, and very different from other fungi. This review has attempted to explain Glomeromycota diversity within the paradigms of population genetics and evolutionary biology, and there are no apparent features of arbuscular mycorrhizal fungi that contradict our understanding of speciation and population differentiation. The most important feature of the Glomeromycota is the possible existence of ancient asexual lineages. Coyne & Orr (2004) suggest that asexual species are not biological species but a collection of 'microspecies', with each individual propagating its own genetic lineage. This certainly seems to be the case with the 'clones' of *Glomus* spp. and the *G. intraradices* isolates from Switzerland (Koch *et al.*, 2006), but could also be the case further in the genealogical hierarchy if the morphological species have evolved asexually. The Glomeromycota species may be recognized as microspecies *sensu* Coyne & Orr (2004), but a significant difference from other such clonally propagated entities is the considerable age of the lineages. The speciation events can be ancient, and even genera within the families may have evolved asexually. If the last meiotic events are ancient, Glomeromycota may consist of very few clones or clonal lineages with several morphological species originating from the same clone. Future studies should use coalescence-based models to determine the age of the speciation events. This is, however, complicated by the ancient origin and asexual evolution with many selective sweeps that have eliminated intermediate lineages. Before any conclusion can be drawn on speciation and population divergence, it is important to note that very few fungi have been studied at the population level: only two studies use population genetics, and these are restricted to agricultural fields in Denmark and Switzerland.

Ectomycorrhiza

Ectomycorrhizal Fungi are, economically, one of the most important groups of fungi. These are the fungi that form a symbiotic relationship with a plant forming a sheath around the root tip of the plant. The fungus then forms a Hartig Net which means that there is an inward growth of hyphae (fungal cell growth form) which penetrates the plant root structure. There are actually seven types of mycorrhiza and 90% of plants form mycorrhiza with fungi, but ectomycorrhizal refers to this sheath forming type.

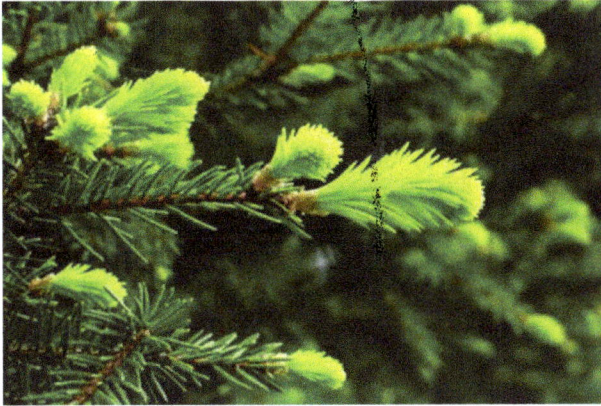

Ectomycorrhizas, or EcM, are typically formed between the roots of around 10% of plant families, mostly woody plants including the birch, dipterocarp, eucalyptus, oak, pine, and rose families and fungi belonging to the Basidiomycota, Ascomycota, and Zygomycota. Ectomycorrhizas consist of a hyphal sheath, or mantle, covering the root tip and a hartig net of hyphae surrounding the plant cells within the root cortex. In some cases the hyphae may also penetrate the plant cells, in which case the mycorrhiza is called an ectendomycorrhiza. Outside the root, the fungal mycelium forms an extensive network within the soil and leaf litter. Nutrients can be shown to move between different plants through the fungal network (sometimes called the wood wide web). Carbon has been shown to move from birch trees into fir trees thereby promoting succession in ecosystems.

The fungi are unique in having a simultaneous dual life-style, living both within the plant roots as symbionts and, at the same time, in the soil as facultative, transitory saprotrophs. Without this symbiosis, forests as we know them could probably not exist because of the essential role of the fungi for tree growth and in the cycling of essential nutrients. Thus, ECM fungi are responsible for a symbiosis of global ecological and economic importance. The trees "feed" the ECM fungus with plant-derived carbohydrates and the fungus then utilizes this energy to decompose and assimilate essential nitrogen and phosphate compounds in the soil and transfer them back to the trees. Moreover, the association of different plants with the same guild of fungi mediates indirect inter-plant interactions, such as nutrient transfer or competition. Understanding how the fungus can achieve this is essential because of the key role of forests in buffering/ sequestering increased CO_2 and also for understanding how to optimize tree productivity in future biofuel production.

The Biology of the Ectomycorrhizal Symbiosis

Fruiting body of the symbiotic fungus Amanita muscaria – The Agaric Fly

Within days after their emergence in the upper soil profiles (e.g. organic humus and mor layer), up to 95% of the short roots of most conifers and deciduous trees are colonised by ectomycorrhizal mycobionts. The approx. 8,000 ectomycorrhizal basidiomycetes (e.g. agarics, boletes) and ascomycetes (e.g. truffles) are not a phylogenetically distinct group, but an assemblage of very different fungal species that have independently developed a symbiotic lifestyle over the last 150-180 millions years. The switch between saprotrophic and mycorrhizal lifestyles probably happened convergently, and perhaps many times, during evolution of these fungal lineages. This may have facilitated evolution of ectomycorrhizal lineages with a broad range of physiological and ecological functions reflecting partly the activities of their disparate saprotrophic ancestors. Whereas a few Basidiomycota clades are exclusively ectomycorrhizal (e.g. bolets), most clades comprise both ectomycorrhizal and saprobic species (e.g. Tricholomataceae) suggesting that the symbiotic ability involved a limited number of (symbiotic) genes.

Although it was thought that ectomycorrhizal fungi were colonizing only trees, the use of DNA-based genotyping methods has recently modified our understanding of the specificity of these fungi towards their host plants. For example, Rhizoctonia spp were reported until recently as the dominant mycorrhizal symbionts of orchids, but direct amplification of fungal DNA from mycorrhizal roots of achlorophyllous and green forest orchids demonstrated that the main symbionts are unculturable fungi that belong to known ectomycorrhizal taxa (truffles, Sebacinaceae). Therefore, the current classification of mycorrhizal associations does not take in account the plasticity of the different fungal groups in forming ectomycorrhizal interactions.

Ectomycorrhiza Morphogenesis

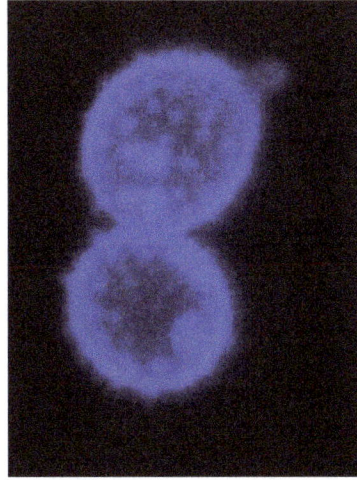

Spores of the symbiotic fungus Laccaria bicolor

Signalling processes bring the mycobiont into the vicinity of susceptible host roots. Only the broad outlines of the signalling processes have been defined, and a limited set of chemical signals produced by either the host or symbiont have been identified so far. Early stages of ectomycorrhiza development have well-characterized similar morphological transitions. Symbiosis development proceeds through a programmed series of morphogenetic events. Fungal hyphae emerging from soil propagules (spores, sclerotia) or an older mycorrhiza penetrate into the root cap cells and grow through them. Backwards from the tip the invasion of root cap cells proceeds inwards until the hyphae reach the epidermal cells. Morphogenetic changes take place upon contact of the hyphæ with living cortical cells, which is pivotal for initiation of mantle formation and Hartig net construction. The hyphae progressing in the root apoplastic space proliferate leading to the formation of a finger-like, labyrinthine system, the so-called Hartig net. Abundant membranes of this structure allow ions, metabolites and effector molecules to pass at a high rate between adjacent cells, providing the anatomical basis for intercellular communication and the local coordination between the symbionts. Numerous mitochondria, lipid bodies, dictyosomes with proliferating cisternae, and extensive endoplasmic reticulum are contained within the coenocytic hyphæ of the Hartig net, all of which are illustrations of a highly active anabolic state with high biosynthesis of secreted proteins. Progression from the strongly rhizomorphic outgrowth of the free-living mycelium to the plectenchymatous structure of the ectomycorrhizal sheath and the coenocytic Hartig net hyphæ is associated with a lack of septation, a loss of apical coherence and intimate juxtaposition of hyphæ. No root cell penetration is observed, except in senescing ectomycorrhizal tips. After attachment onto epidermal cells, hyphae multiply to form a series of layers of several hundred μm thick which differentiates to form the mature mantle. The hyphae in these structures are encased in an extracellular polysaccharide and proteinacous matrix. Air and water channels that allow the flow of nutrients into the symbiosis innervate these structures, although most of the nutrient transfer probably takes place via the symplastic way. An outward network of hyphae prospecting the soil and gathering nutrients irradiate from the outer layers of the mantle.

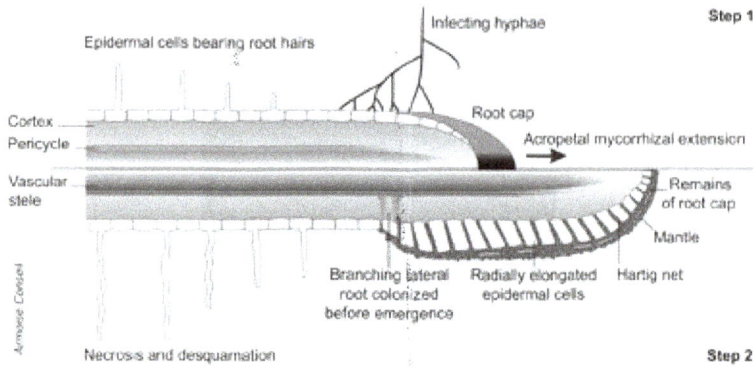

Ecology

Biogeography and Environmental Gradients

Ectomycorrhizal fungi are found throughout boreal, temperate and tropical ecosystems, primarily among the dominant woody-plant-producing families. In these more mesic environments supporting coniferous and mixed coniferous and deciduous forests, the EcM produce proteases and acid phosphatase enzymes to access organic forms of both nitrogen and phosphorus.

Many of the fungal families most common in temperate forests (e.g. Russulaceae, Boletaceae, Thelephoraceae) are also quite widespread in the southern hemisphere and tropical dipterocarp forests. While there are differences between the fungal makeup of these different ecosystems, the ectomycorrhizal fungal component shows much greater similarities than the minimal overlap that occurs between dominant plant families in temperate and tropical forests.

There is evidence to suggest that communities of EcM fungi differ across soil type gradients in a tropical system. However, the particular study notes that the mechanism driving this differentiation is not clear, and the variance could be in response to the soil physiochemical environment, plant community, or both. Other studies offer evidence to strengthen the idea that EcM communities are indeed affected by soil environment both in the field and in the lab.

Some studies indicate that ectomycorrhizal fungi might be at odds with the general latitudinal gradient of diversity (LGD). Data sets, free from inconclusive fruit-body surveys and largely relying on more accurate sequencing and microarray technologies, indicate that EcM fungi may be at enhanced diversity in the temperate zone. This implies that many of the causal mechanisms proposed to explain the LGD pattern might be inapplicable, or in need of modification, in reference to EcM. Though this relationship is far from certain, there exist some hypotheses to explain the plausibility of this phenomenon: 1) EcM fungi may have evolved at higher latitudes with Pinaceae hosts, and are subsequently inferior at competition in tropical climates, 2) host lineages might be more diverse in temperate conditions, and well developed soil and soil horizons in temperate regions allow for higher niche differentiation and species accumulation, and 3) tropical EcM hosts are more sparsely distributed, yielding small isolated forest islands that may reduce the population sizes and subsequent richness of EcM fungi.

This third hypothesis is expounded on in a study demonstrating that habitat size also plays an important role in determining the species richness and assemblage structure of ectomycorrhizal fungi, namely that richness is reduced in smaller and more isolated habitat areas. The same study determined that spatial turnover of soil fungi actually occurs on scales more similar to macro-organisms.

Similar to studies concerning nitrogen eutrophication, EcM fungal makeup over an anthropogenic nitrogen gradient show similar trends. Species richness declined dramatically with increasing nitrogen inputs, with over 30 species represented at low nitrogen sites and only 9 at high nitrogen sites. It is speculated that as nitrogen increases, taxa shift from those specialized for low nitrogen conditions to those specialized for phosphorus uptake in low phosphorus, high nitrogen, acidified conditions.

Host Specificity and Community Responses

Across most EcM host lineages, there appears to be low levels of specificity, as EcM plants tend to form symbioses with many distantly related fungi. The benefits of this system are twofold: 1) seedlings are more likely to form mycorrhizas in a wide array of habitats, thus extending range and setting, and 2) EcM mycobionts can differ in their ability to access nutrients, thus allowing host plants better access to these minerals. Many species of *Alnus* exhibit a very narrow range of fungal symbionts, but these fungi are not from related lineages. Thus it is the mycobionts that show phylogenetic specificity, not the alders.

A major exception to the general rule outlined above is exemplified by mycoheterotrophic plants that utilize ectomycorrhizas for their carbon needs. These plants, from the subfamily Monotropoideae, exhibit high specificity for the EcM fungi they parasitize, although different monotrope species target a rather wide array of fungal lineages.

While the plant hosts exhibit low specificity, EcM fungi exhibit various levels of specificity, and the costs and benefits to their specialization are not well understood. A good example is the suilloid group, a monophyletic assemblage containing the genera *Suillus*, *Rhizopogon*, *Gomphidius* and others. This is the largest group that exhibits such an extreme degree of specificity, with almost all of its members forming ectomycorrhizas with members of the Pinaceae. However, many other fungal groups exhibit a very broad host-range more akin to host lineages.

Host plants that are taxonomically related show more similar EcM fungal communities than do taxa that are more distantly related. Molecular phylogenetic studies have
shown that fungi derived from a common ancestor are more likely to show host specificity to plants that are taxonomically related. Host successional status may also play a
role in determining EcM fungal communities, as well as affecting the number of EcM
fungal species associated with an individual host species. Other indirect factors can
also play a role in the EcM fungal community, such as leaf fall and litter quality, which
subsequently affect calcium levels and soil pH. Even establishment timeframe of the
host species can have an effect, with lower EcM fungal richness associated with hosts
from a secondary forest than from a primary forest.

The establishment of common mycelial networks is thought to have effects upon the
plant community involved with them through their ectomycorrhizal connections. This
can range from providing access to larger nutrient pools, mediating competition, and
allowing resources and nutrients to be shared among individuals linked in this manner.

Roles in Invasion

Pine plantation, probably inoculated with fungal spores
to allow beneficial ectomycorrhizas to form

Mycorrhizas have been regarded as the most prevalent symbiotic condition on earth,
and as such they are essential to plant nutrition in terrestrial ecosystems. Thus, even
alien plants often require mycorrhizal symbionts for the establishment and spread into
foreign environments. Due to the low specificity of the vast majority of arbuscular mycorrhizas, AM plants often become invasive quickly and easily, and as such, the invasions are not necessarily accompanied by a simultaneous AM fungal invasion. However,
because ectomycorrhizal symbioses present a range of specificities, exotic forestry has
often relied upon the introduction of compatible EcM fungi to the foreign landscape in
order to ensure the success of forest plantations and the like.

This is most common in eucalypts and pines, which are obligate ectomycorrhizal trees
in natural conditions. This is evidenced by the struggle of establishment of pines in
the southern hemisphere until the anthropogenic buildup of soil inoculums. Similarly,

Australian eucalypts and acacias have evolved in isolation from the EcM fungi associated with many other temperate trees such as *Pinus* and *Quercus*. Thus, much like pines in the southern hemisphere, many *Eucalyptus* plantations required inoculation by EcM fungi from their native landscape. In both cases, EcM networks allowed for the naturalization of the introduced species, followed quickly by competition for resources with native plants and invasion into novel habitats.

Many EcM species co-invade without the help of human activity, however. Members of Pinaceae represent another prime example of this convention, often invading habitats along with specific EcM fungi from the genera *Suillus* and *Rhizopogon*. There are, however, ectomycorrhiza-forming fungi with cosmopolitan distributions. These EcM fungi allow non-native plant species to form mutualisms that are not novel in environments that are, thus bypassing the need for co-invasion with specific EcM fungi from the native ecosystem.

Dominant native plants are capable of inhibition of EcM fungi on the roots of neighboring plants through the release of chemical compounds or through competitive interactions. Some invasive plants are capable of inhibiting the growth of native ectomycorrhizal fungi through similar mechanisms, especially if they become established and dominant. Invasive garlic mustard, *Alliaria petiolata*, and its allelochemical benzyl isothiocyanate were shown to inhibit the growth of three species of EcM fungi grown on white pine (*Pinus strobus*) seedlings. Changes in EcM communities can have drastic effects on nutrient uptake and community composition of native trees, which can in turn have far-reaching ecological ramifications.

Competition and other Plant Symbionts

Competition among EcM fungi is a well-documented case of soil microbial interactions. In many experimental cases, the timing of colonization between competing EcM fungi determined which species was dominant. Namely, there was a priority effect that significantly favored the original colonists to be the most dominant, except in cases involving fungal species at a natural competitive disadvantage. This disadvantage appears to be related to the proportion of root tips colonized, and those species incapable of colonizing a sufficient proportion of host roots do not typify this priority effect.

Many other biotic and abiotic factors can mediate competition among EcM fungi, such as temperature, soil pH, soil moisture, host specificity, and competitor number. The results of many studies concerning these factors indicate that these interactions are largely environmentally context-dependent. These aspects can often lead to "checkerboard" distribution patterns, where certain species occupy locations that are mutually exclusive of the other species.

EcM communities continue to exhibit rare EcM fungal constituents that have not been excluded, despite intense competition. Thus, mechanisms must exist that maintain diverse levels of EcM fungi. This coexistence can be summed up in four non-mutually exclusive possibilities illustrated by Bruns: niche partitioning, disturbance-related patch dynamics, density-dependent mortality and competitive networks.

There is also some evidence for competition between EcM fungi and arbuscular mycorrhizal fungi. This is mostly noted in species, such as certain eucalypts, that are capable of hosting both EcM and AM fungi on their roots. There is also some evidence in larger scale systems, such as pinyon woodlands, although it is hard to extricate effects of mycorrhizal interactions (if there are any) from those of simple resource competition.

Some soil bacteria have been shown to have beneficial effects upon the establishment of ectomycorrhizal symbioses. Some of these bacteria, known as Mycorrhiza helper bacteria (MHBs), have been shown to stimulate EcM formation, root and shoot biomass. The presence of higher levels of ergosterol in the soil indicate that MHBs may be promoting fungal growth, as well, thereby generating an increase in mycelial tissue and hyphae capable of exploring greater soil volumes. The mechanisms by which these bacteria stimulate mycorrhizal formation are unclear. However, some mechanistic hypotheses include the softening of cell walls to make root cells more receptive, stimulation of short root formation in plants to allow for a higher probability of encounters with fungal propagules, and mediation of chemical elicitors involved in mutual recognition. However, regardless of mechanism, it is becoming evident that bacteria are more ubiquitous than previously thought and could represent a third component of mycorrhizas (3.4.l).

However, not all bacteria exhibit beneficial effects, and there exists some bacteria whose effects are quite opposite to those of MHBs, thus inhibiting ectomycorrhizal formation.

Faunal Interactions

The edible epigeous fruiting body of *Cantharellus cibarius*, the golden chanterelle

Many ectomycorrhizal fungi are known to rely upon mammals for the dispersal of spores, particularly those fungi with hypogeous fruiting bodies. Many species of small mammals exhibit a high degree of mycophagy, ingesting a wide taxonomic range of fungi. These mammals are often drawn to hypogeous fruiting bodies because they are rich in nutrients such as nitrogen, phosphorus, minerals and vitamins. However, some sources state these nutritive properties are overstated, and it is more likely due to availability at specific times of the year, ease of harvest, and patchy nature of distribution.

The spores of these fungi are dispersed either by the actions of being unearthed and broken apart, or by ingestion and subsequent excretion. Some studies even suggest that passage through an animal's gut promotes the germination of these spores, although it is by no means necessary for a majority of fungal species. Regardless, the ability of these certain mammals to spread fungal spores is thus indirectly related to plant community structure, by way of the pivotal role that EcM fungi play in plant nutrition and productivity.

Many other sporocarps are grazed upon by invertebrates such as mollusks and fly larvae, some of which are even tolerant to the toxic α-amanitin. Belowground, populations of nematodes and springtails are maintined by consumption of fungal tissue. There are also interesting studies concerning EcM fungi and arthropods. The ectomycorrhizal fungus *Laccaria bicolor* has been found to lure and kill springtails to obtain nitrogen, some of which may then be transferred to the mycorrhizal host plant. In a study by Klironomos and Hart, eastern white pine inoculated with *L. bicolor* was able to derive up to 25% of its nitrogen from springtails.

Of course, edible fungus plays a role in many societies throughout the world, as well. Many epigeous mushrooms are collected and consumed on a regular basis, and more recent commercial harvesting is beginning to play a larger economic role in certain locales. Certainly, truffles (*Tuber*), porcinis (*Boletus*) and chanterelles (*Cantharellus*) are commonly known for their taste and culinary importance, as well as their billion dollar worldwide market.

Plant Production

Agriculture

Ectomycorrhizal fungi do not play a large role in agricultural and horticultural systems. Most of the economically relevant crop plants that form mycorrhizas tend to form them with arbuscular mycorrhizal fungi. Many modern agricultural practices such as tillage, heavy fertilizers, and fungicides have extremely detrimental effects on crops' associated mycorrhizas and on the surrounding ecosystem. Thus, it is possible that agriculture indirectly affects nearby ectomycorrhizal species and habitats, such as increased fertilization decreasing sporocarp production.

Forestry

In commercial forestry, the transplanting of crop trees in new locales often requires an accompanying ectomycorrhizal partner. This is especially true of trees that have a high degree of specificity for their mycobiont, or trees that are being planted far from their native habitat among novel fungal species. This has been shown time and again in plantations involving obligate ectomycorrhizal trees, such as *Eucalyptus* and *Pinus* species. Mass planting of these species often require human addition of inoculum from native EcM fungi in order for the trees to prosper.

Thus, these EcM fungi have to be species that are capable of being grown in bulk. After being added to various soil mixtures, the mutualism can begin as seedlings are grown in nurseries or plantations. This is already becoming quite commonplace, and there are many companies that are beginning to sell a variety of mycorrhizal inoculum, *Pisolithus tinctorius* being quite widespread among the EcM fungi.

Sometimes ectomycorrhizal plantation species, such as pine and eucalyptus, are planted and promoted for their ability to act as a sink for atmospheric carbon. However, the ectomycorrhizal fungi of these species also tend to deplete soil carbon over relatively short periods of time. Thus there is a great deal of mounting resistance to using tree plantations as general solutions to combatting rising carbon dioxide levels.

Restoration

Ectomycorrhizas provide many benefits to their host plants, with enhanced nutrient uptake, growth and establishment in disturbed habitats ranked highly among them. Thus, it seems logical that EcM fungi could be used in restoration projects aimed at re-establishing native plant species in ecosystems disrupted by a variety of issues. In addition to providing a certain degree of protection to seedlings in harsh circumstances, such as increased salinity or heavy metal pollution, the fungi are also instrumental in improving soil quality. They are able to achieve this through allowing the establishment of early vegetation and subsequent organic litter, preventing erosion, and binding soil particles together yielding stability and soil aggregation. Since the disappearance of mycorhizal fungi from a habitat constitutes a major soil disturbance event, its re-addition is an important part of establishing vegetation and restoring habitats.

Climate changes have important consequences for plant communities and their root symbionts. The distribution of tree species within temperate, boreal and tropical biomes will be altered, as palaeoecological studies have demonstrated for previous climate change events. Predicted effects on ectomycorrhizal (ECM) associations include migration of host and symbiont, modification of interactions between plant and fungal species, and changes in the contribution of both partners to the global carbon cycle. Anthropogenic factors introduce new variables, affecting the ability of tree species and their fungal associates to disperse in response to climate change. Here we focus on how ECM fungi and their hosts respond to atmospheric CO_2 enrichment, increasing temperatures, nutrient addition, species invasions, loss of biodiversity and anthropogenic land-use changes, particularly silviculture. All of these factors are key to understanding the impacts of climate change on the ECM symbiosis, and relevant future topics of research are presented.

▶ ECM fungi respond variably to climate change. ▶ Palaeoecology indicates symbiosis survived many cycles of natural climatic change. ▶ Fungal respiration and carbon storage significantly impact the carbon cycle. ▶ Human land-use changes can interfere with host and symbiont dispersal. ▶ ECM-dominated ecosystems require management for complexity.

Ericoid Mycorrhiza

Ericoid mycorrhiza (ERM) is a symbiotic association between fungi and the roots of plants in the families Ericaceae and Diapensiaceae (e.g. Schizocodon and Diapensia) .The structure of ERM is characterized by hyphal coils formed in the epidermal cells of the extremely thin "hair roots" of hosts. While ERM is widespread in temperate regions, it is particularly dominant in the dwarf shrub vegetation of alpine and arctic regions. Mineralization and decomposition rates in these regions are extremely low due to low temperatures, and, thus, most of the nitrogen in soil exists in organic forms that are unavailable to most plants (33). ERM host plants have the ability to utilize many forms of organic nitrogen because ERM fungi have the capacity to decompose organic matter .Therefore, forming a relationship with ERM fungi is essential to the success of ERM host plants, such as dwarf shrubs of the family Ericaceae, in alpine and arctic regions .

Fungal Symbionts

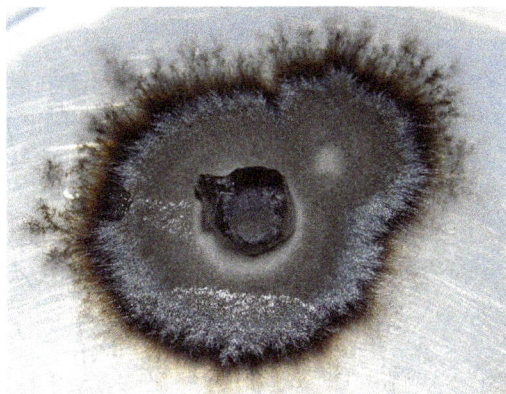

An isolate of the ericoid mycorrhizal fungus, *Gamarada debralockiae*, isolated from *Woollsia pungens*

The majority of research with ericoid mycorrhizal fungal physiology and function has focused on fungal isolates morphologically identified as *Rhizoscyphus ericae*, in the Ascomycota order Helotiales, now known to be a *Pezoloma* species.

In addition to *Rhizoscyphus ericae*, it is currently recognized that culturable Ascomycota such as *Meliniomyces* (closely allied with *Rhizoscyphus ericae*), *Cairneyella variabilis*, *Gamarada debralockiae* and *Oidiodendron maius* form ericoid mycorrhizas.

The application of DNA sequencing to fungal isolates and clones from environmental PCR has uncovered diverse fungal communities in ericoid roots, however, the ability of these fungi to form typical ericoid mycorrhizal coils has not been verified and some may be non-mycorrhizal endophytes, saprobes or parasites.

In addition to ascomycetes *Sebacina* species in the phylum Basidiomycota are also recognized as frequent, but unculturable, associates of ericoid roots, and can form ericoid mycorrhizas. Similarly, basidiomycetes from the order Hymenochaetales have also been implicated in ericoid mycorrhizal formation.

Geographic and Host Distribution

The ericoid mycorrhizal symbiosis is widespread. Ericaceae species occupy at least some habitats on all continents except Antarctica. A few lineages within the Ericaceae do not form ericoid mycorrhizas, and instead form other types of mycorrhizas, including manzanita (*Arctostaphylos*), madrone (*Arbutus*), and the Monotropoidiae. The geographic distribution of many of the fungi is uncertain, primarily because the identification of the fungal partners has not always been easy, especially prior to the application of DNA-based identification methods. Fungi ascribed to *Rhizoscyphus ericae* have been identified from Northern and Southern Hemisphere habitats, but these are not likely all the same species. Some studies have also shown that fungal communities colonizing ericoid roots can lack specificity for different species of ericoid plant, suggesting that at least some of these fungi have a broad host range.

Economic Significance

Ericoid mycorrhizal fungi form symbioses with several crop and ornamental species, such as blueberries, cranberries and *Rhododendron*. Inoculation with ericoid mycorrhizal fungi can influence plant growth and nutrient uptake. However, much less agricultural and horticultural research has been conducted with ericoid mycorrhizal fungi relative to arbuscular mycorrhizal and ectomycorrhizal fungi.

Orchid Mycorrhiza

Orchid Mycorrhizae refers to the symbiotic relationships between the roots of the plants of the Orchidaceae and a variety of fungi.

Orchid mycorrhizae may be considered the epitome of plant fungal interdependency in the mycorrhizal world. Some 17 000 species of orchids exist and depend upon their basidiomycete fungal partners for acquisition of nutrients. In some cases, the dependency of the plant on its fungal partner has become so extreme that fungal propagules are carrier within the seed of the orchid in order to be present at seed germination and enhancing nutrient uptake at an early stage of plant development. This is important where seed size is so small as to limit the nutrient reserve that can be carried in the absence of an endosperm. Initial protocorm development after seed germination is dependent on the symbiotic fungus also for carbohydrate supply, until photosynthetic capacity can be developed. This may take up to a year in some species, and in achlorophyllous orchid species, it never develops. The fungi develop highly coiled arbuscules (peletons) within the cortical cells of the host plant. These fungal coils have a finite lifespan and upon death, deposit cellulose and pectin within the host cell. These cells may be subsequently 'invaded' by new hyphae that can access these carbohydrate and nutrient supplies.

Fungal Entry into Orchid

The fungus can enter at various orchid life stages. Fungal hyphae can penetrate the parenchyma cells of geminated orchid seeds, protocorms, late-staged seedlings, or adult plant roots. The fungal hyphae that enter the orchid have many mitochondria and few vacuoles. In the protocorm stage, hyphae enter the chalazal (top) end of the embryo. In terrestrial orchids, fungal entry into adult plant roots happens mainly through root hair tips, which then take on a distorted shape. Typically, the partnership is maintained throughout the lifetime of the orchid because they depend on the fungus for nutrients including minerals. However, some orchids have been found to switch fungal partners during extreme conditions.

Fungal Pelotons and Orchid Root Cortex

Shortly after the fungus enters an orchid, the fungus produces intracellular hyphal coils, called pelotons, in the embryos of developing seedlings and the roots of adult plants. The formation of pelotons in root cortical cells is a defining anatomical structure in orchid mycorrhiza that differentiate it from other forms of mycorrhiza. The pelotons can range in size and in the loose or tight packaging of their hyphae. Pelotons of live fungal hyphae are eventually disintegrated, or lysed, and become brown or yellow clumps in the orchid cells. The disintegrated pelotons are an area of considerable interest in current research. The disintegrated pelotons first experience a collapse where orchid microtubules surround the pelotons, which may be the mechanism behind the peloton collapse by producing physiological and structural changes of the hyphae. The cortical cells of older roots tend to have more lysed pelotons than young pelotons. Although pelotons are lysed, new pelotons continue to be formed, which indicates a high amount of entering hyphal activity. Pelotons are separated from the orchid's cytoplasm by an interfacial matrix and the orchid's plasma membrane. The material that makes up the interfacial matrix can differ depending on the orchid mychorrizal stage of interaction. Orchid cells with degenerating pelotons lack starch grains, whereas the newly invaded orchid cells contain large starch grains, suggesting the hydrolysis of starch resulting from the fungal colonization. There is an enlargement of the nucleus in infected orchid cortical cells and in non-infected cortical cells near an infected area as a result of increased DNA content. The increased DNA content has been correlated with the differentiation of parenchyma cells suggesting its role in orchid growth.

Fungi forming Orchid Mycorrhizae

The fungi that form orchid mycorrhizae are typically basidiomycetes. These fungi come from a range of taxa including *Ceratobasidium* (Rhizoctonia), *Sebacina*, *Tulasnella* and *Russula* species. Most orchids associate with saprotrophic or pathogenic fungi, while a few associate with ectomycorrhizal fungal species. These latter associations are often called tripartite associations as they involve the orchid, the ectomycorrhizal fungus and its photosynthetic host plant. Some of the challenges in determining host-specificity in orchid mycorrhizae have been the methods of identifying the orchid-specific fungi from other free living fungal species in wild-sourced samples. Even with modern molecular analysis and genomic databases, this can still prove difficult, partially due to the difficulty in culturing fungi from protocorms and identification of fungal samples, as well as changes in evolving rDNA. However it has become clearer that different fungi may associate with orchids at specific stages, whether at germination or protocorm development, or throughout their life. The types of orchids and their symbiotic fungi also vary depending on the environmental niches they occupy, whether terrestrial or growing on other plants as an epiphyte.

Symbiont Specificity

Current molecular analysis has allowed for the identification of specific taxa forming symbiotic relationships which are of interest in the study, cultivation, and conservation of orchids. This is especially important in the trade and preservation endangered species or orchids of commercial value like the vanilla bean. There have been seen trends in the type of symbioses found in orchids, depending primarily on the life-style of the orchid, as the symbiosis is primarily of benefit to the plant. Terrestrial orchids have been found to commonly associate with Tulasnellaceae, however some autotrophic and non-autotrophic orchids do associate with several ectomycorrhizal fungi. Epiphytic fungi, however, may associate more commonly with limited clades of *rhizoctonia*, a polyphyletic grouping. These fungi may form significant symbioses with either an epiphytic or terrestrial orchid, but rarely do they associate with both. Using seed-baiting techniques researchers have been able to isolate specific species and strains of symbiotic orchid mycorrhizae. Using this technique seeds of the epiphytic orchid *Dendrobium aphyllum* were found to germinate to seedlings when paired with fungal strains of *Tulasnella*, however, seeds of this species did not germinate when treated with *Trichoderma* fungi taken from adult orchids, indicating there are stage specific symbionts. These fungal symbioses, as well as their affinity towards specific symbionts, vary based on the stage of development and age of the host roots As an orchid ages the fungal associations become more complex.

Mixotrophic orchids like *Cephalanthera longibracteata* may associate generally with several fungi, most notably from *Russulaceae,Tricholomataceae, Sebacinales, Thelephoraceae.*, as they do not depend as heavily on the fungus for carbon. Some orchids can be very specific in their symbionts, preferring a single class or genus of fungi. Genotypes of *Corallorhiza maculata,* a myco-heterotrophic orchid, have been found to closely associate with *Russulaceae* regardless of geological location or the presence of other orchids. The terrestrial Chilean orchids *Chloraea collicensis* and *C. gavilu* have been found to have only one key symbiont in the genus *Rhizoctonia*. Research suggests that orchids which are considered to be generalists will associate with other generalist fungi, often with several species, however orchids with high degrees of specialization will have less fungal associations.

The environment may also affect the fungal symbiosis, with differences in tropical versus temperate orchids. The photosynthetic orchid *Goodyera pubescens* was found to associate with only one dominate fungus, unless subjected to changes in the environment, like drought, in which case the orchid was able to change symbionts in response to stresses. This is unique in that it conflicts with the trends seen in most arbuscular and ectomycorrhizal symbioses which form associations with several fungi at the same time.

The knowledge of species-specific fungal symbionts is crucial given the endangered status of many orchids around the world. The identification of their fungal symbionts could prove useful in the preservation and reintroduction of threatened orchids. Researchers in India have used samples from adult orchids to germinate seeds from an endangered orchid *Dactylorhiza hatagirea* which were found to associate closely with *Ceratobasidium*.

Functions of Mycorrhizae

This relationship between the fungus and a plant can obviously have a great impact on the environment. In areas where drought is prevalent, the plants that are able to use mychorrhizae to increase root surface area and obtain water will have an advantage over those without this symbiotic relationship. The same can be said about environments that are low in nutrients such as nitrogen and phosphorus. The plants that can have this association with mycorrhizal fungi will have a greater chance in inhabiting this area.

Nutrient Acquisition

The major function of mycorrhizae is their ability to exchange nutrients between their surroundings and their host plant. With the increase in root surface area and the protection they offer to the plant's roots, the fungus is able to acquire a lot of nutrients for its host. Mycorrhizae are able to uptake water, inorganic phosphorus, mineral or organic nitrogen, and amino acids through specialized transporters located on their membrane (3). Once the water and nutrients are acquired, they can then be transferred to the host, who in return supplies carbon.

Aiding in Agriculture

Since mycorrhizal relationships can help to increase plant growth and therefore yield, they can be beneficial in agricultural fields. An increase in yield for farmers also means an increase in income. This symbiotic relationship in agricultural fields would increase crop production and therefore increase food output.

The mycorrhizal fungi also aids in soil aggregation, which can increase water filtration and gas exchange within the soil (7). With an increase in gas exchange, the mycorrhizal fungi can aid in the aeration of agricultural fields. This, along with the other benefits of mycorrhizae, can increase the crop yield.

Aiding in Restoration

This symbiotic relationship can also be helpful in restoration areas. Since the fungus allows certain plants to be greater competitors and can allow them to prosper in areas low in nutrients and water, they would be very useful in restoration efforts. Not only could this partnership increase a plant's ability to colonize an area, they would also have a greater capacity to out compete.

Current Research

Since mycorrhizae can form beneficial relationships with plants, many experiments have been performed to research the extent of these advantages.

"Influence of Arbuscular Mycorrhizae on the Root System of Maize Plants under Salt Stress"

This experiment aimed at discovering the effect of salinity on corn with mycorrhizal relationships. The effect of salinity was tested because of its ability to reduce growth and yield of crops. On the other hand, arbuscular mycorrhizae (endomycorrhizae) is known for its ability to increase the growth and yields of plants. In order to conduct this experiment, corn was planted in soil with five different salinity levels for a total of fifty-five days. Half the corn in each different salinity level had a mycorrhizal partnership while the other half of the corn did not. After the days were up, the plants were then removed and their root biomass, root morphology, and root activity. The compiled data showed that the corn associated with the arbuscular mycorrhizae had a larger root biomass, root morphology, and root activity in all of the salinity levels compared with the corn without the symbiotic association. All in all, this research showed that the corn with the arbuscular mycorrhizae was able to alleviate some of the stress caused by high salinity levels.

"Contribution of Mycorrhizae to Early Growth and Phosphorus Uptake by a Neotropical Palm"

In this experiment, the association between *Desmoncus orthacanthos* and arbuscular mycorrhizae was tested. *Desmoncus orthacanthos* is known for its large roots that do not branch and do not have many hairs. Due to this, research was done to test the ability of the mycorrhizae to increase phosphorus uptake and to increase the growth of the plant seedlings. Seedlings were planted in soil with three different levels of phosphorus for a total of one hundred and sixty days. Half of the seedlings had a partnership with the arbuscular mycorrhizae while the other half did not have an association with this fungus. After the experiment was over, the growth of each seedling was measured. Overall, the concentration of phosphorus in the seedlings increased with the mycorrhizae association as well as with the increased phosphorus in the soil. This experiment revealed that it is very beneficial for *Desmoncus orthacanthos* seedlings to have a partnership with arbuscular mycorrhizae especially when they are in soil with low levels of phosphorus.

"Arbuscular Mycorrhizae Improves Low Temperature Stress in Maize via Alterations in Host Water Status and Photosynthesis"

This experiment tested the effect of low temperatures on the growth of corn. Low temperatures are an abiotic factor that can greatly reduce and limit the growth of plants. With this said, this experiment tested the ability of arbuscular mycorrhizae on the water uptake, growth, and chlorophyll concentration in corn. In order to carry out this research, corn with and without arbuscular mycorrhizae associations were planted in soil for seven weeks in a temperature of twenty-five degrees Celsius. Next, the plants were introduced to temperatures of five degrees Celsius, fifteen degrees Celsius, and twenty-five degrees Celsius for a week. After the results were compiled, it was shown that as the temperature decreased, so did the concentration of arbuscular mycorrhizae on the roots of the corn that had this symbiotic relationship. Although the concentration of the fungi did decrease with temperature, the corn with the mycorrhizal association showed a higher plant growth, dry root weight, water uptake, chlorophyll concentration, and photosynthesis rate than the corn without this partnership. Overall, this experiment came to the conclusion that with arbuscular mycorrhizae, corn has the ability to improving its water uptake and photosynthesis rate when exposed to low temperatures.

Myco-heterotrophy

There are a number of flowering plants that have abandoned photosynthesis, and these fall into two categories: haustorial parasites and mycoheterotrophs. Some choose to call all of these "parasitic plants", and in one sense this is true because both groups derive their nutrients from another plant. The difference is that haustorial parasites feed directly on another plant via a modified root called the haustorium whereas mycoheterotrophs obtain their nutrition indirectly from the plant via a mycorrhizal fungus. The mycorrhizal fungus, attached to the roots of a photosynthetic plant, thus acts as a bridge between that plant and the mycoheterotroph, such that nutrients (carbon) flow from plant root, to mycorrhizal fungus to the mycoheterotroph. One may see these plants called mycoheterotrophic epiparasites or ectomycorrhizal epiparasites. These epithets are quite descriptive (albeit cumbersome) in that the flowering plant could be considered an epiparasite of the fungus. Personally, to avoid confusion, I prefer to use the terms "parasite" or "parasitic plant" when referring to haustorial parasites, and "mycoheterotroph" or "mycoheterotrophic plant" when referring to the other nutritional mode. Of course both groups are heterotrophs, but they obtain their nutrition in different manners. And both groups have mixotrophic (combination of photosynthesis and heterotrophic feeding) and fully-heterotrophic (non-photosynthetic) representatives. Finally, mycoheterotrophs are sometimes mistakenly called saprophytes. There are no true saprophytes in the angiosperms. Only fungi can directly utilize dead organic material.

Biological Interaction

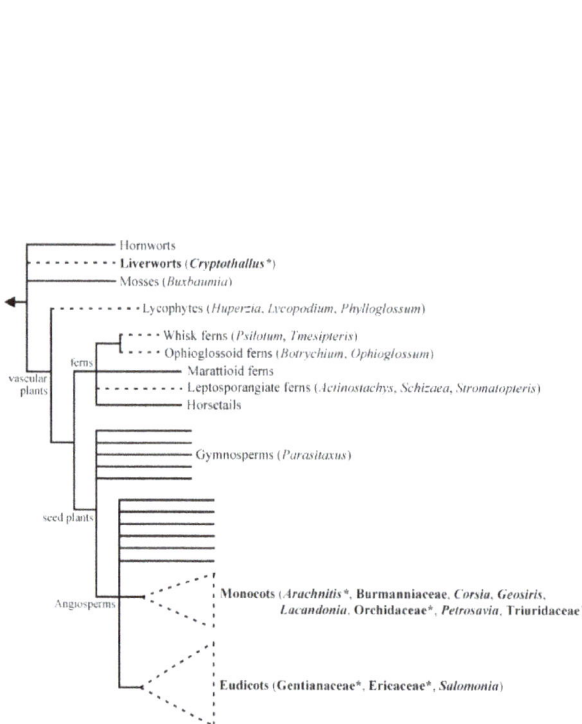

A plant family tree displaying taxa with mycoheterotrophic individuals with a dashed line

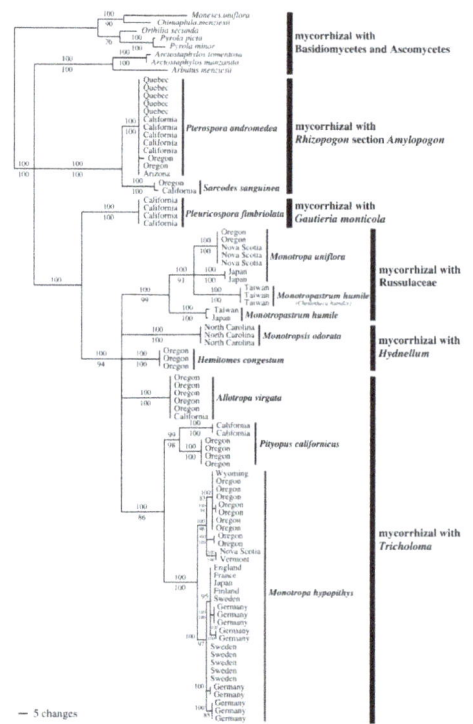

Consensus phylogram of fungal associations of mycoheterotrophic plants in the family Monotropidae from Bidartondo

Mycoheterotrophic plants are very host specific. They associate both with saprotrophic fungi, which obtain carbon from the decomposition of dead plant material, and mycorrhizas that obtain carbon from host-plant photosynthate (Leake 2005). Mycoheterotrophs are often--but not exclusively--associated with patches of ectomycorrhizal plants, which sustain the fungal host with a source of carbon, facilitating the production of a soil mycelial network (McGuire 2007).

These mycelia can interact with mycoheterotrophic roots, which have a specialized root-ball morphology. The plant root-balls are encapsulated by a hyphal sheath similar to those characteristic of the plant-ectomycorrhizal interface. However, the Hartig net, which acts as the source of carbon and nutrient exchange in mutualistic ectomycorrhizas, does not penetrate beyond the mycoheterotrophic epidermis. This morphological difference effects a unidirectionality of carbon flow through the mycelial network, making the mycoheterotroph exclusively a carbon sink.

Considering their total dependence of on fungal carbon, and maximum adult sizes of up to 2 meters tall (*Pterospora andromedea*) and masses of several kilograms (*Sarcodes sanguinea*, mycoheterotrophs must consume a substantial proportion of the photosynthate transferred to ectomycorrhizas. While the effect on both the fungus and ectomycorrhizal host plants should be negative, it is not clear that this is ecologically relevant (Leake 2005).

Mycoheterotrophy can be entirely obligate, or else partial, in which case it is called "Mixotrophy". Plants in these categories can be distinguished by their dependence on fungal inoculation, a morphological basis, and $^{13}C/^{12}C$ isotopic signature (Leake 2005, McGuire 2007.

Obligate Mycoheterotrophy

Obligate mycoheterotrophs are totally acholorophyllous. They obtain 100% of their carbon from their fungal hosts. The order Orchidales contains the most fully mycoheterophic individuals. Similarly the ericaceous subfamily Monotropoideae, Truridaceae, Petrosaviaceae, and Corsiaceae are entirely mycoheterotrophic.

Mixotrophy

Plants that either (1) maintain some stem and/or leaf photosynthetic capacity (partial mycoheterotrophy) or (2) receive supplemental carbon from a common mycelial network (facultative mycoheterotrophy) in order to tolerate the shady understory are called mixotrophs.

Partial Mycoheterotrophy

Many mixotrophs maintain some photosynthetic capacity. Thus, their total carbon acquisition is derived simultaneously from the atmosphere (CO_2) and their fungal host. Several plant families contain genera that represent several different trophic strategies.

Plants in the family Orchidioideae range along a continuous spectra from full autotrophy ("green orchids"), intermediate mixotrophs (including members of the genera *Cephalanthera* and *Epipactis*), to the obligate mycoheterotrophs. Burmanniaceae contains a mix of mycoheterotroph and mixotrophs. Gentianaceae is mostly autotrophic, but with about 30 mycoheterotroph genera (e.g. *Obolaria* and *Bartonia*)

Facultative Mycoheterotorphy

Plants that form a classical ectomycorrhizal colonization can receive supplemental carbon from the mycelial network at early life stages under the plant canopy, and then change to autotrophy in the presence of sunlight. This process is thought to be especially important in maintaing ectomonodominant tree stands--monoculture patches of ectomycorrhizal trees--within tropical forests characterized by a diverse flora of predominately arbuscular mycorrhizal tree species. "Common ectomycorrhizal networks may maintain monodominance in a tropical rainforest".

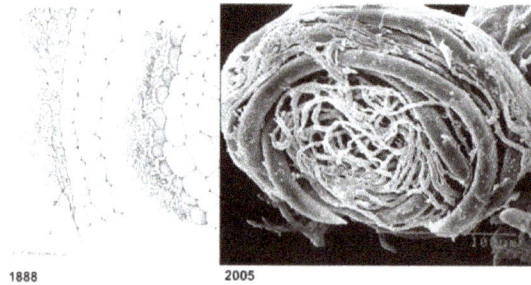

1888   2005

The earliest known drawing of a mycorrhizal fungus (left, from 1888), showing the sheathing of hyphae along the root epidermis and a layer of dead plant cells. An electron micrograph (right) of the infected roots of the mycoheterotrophic *Afrothismisa* from 2005 shows the same thing in three-dimensions.

Root system of *Voyria tenella*, showing the root-ball morphology

Niche

Because mycoheterotrophs are dependent on their fungal hosts for carbon, they tend to be distributed around trees that associate with their mycorrhizal hosts. Also, as their is no competitive advantage for capturing light, they can remain in the shady understory (Leake 2005).

Mycoheterotrophic plants are generally very host specific, with some exceptions (Hynson and Bruns 2009). Germination of mycoheterotrophic plant seeds is stimulated by the presence of host-fungus, such that seeds can enter long-term dormancy if they are isolated from their hosts (Yagame *et al.* 2009). However, very little is known of the chemical signals, or the genetic basis, of the plant-fungal interaction.

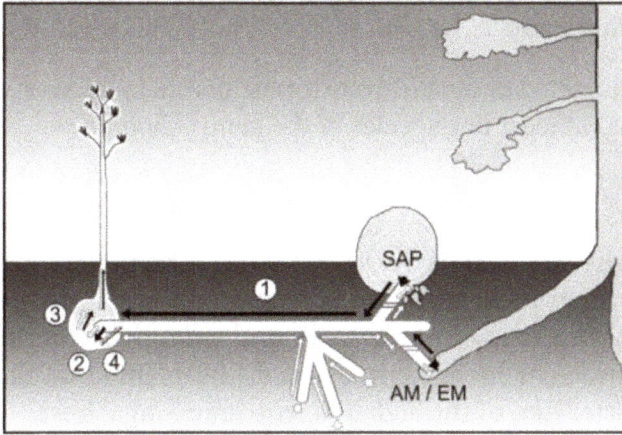

A schematic diagram of the mycoheterotrophic interaction between a plant and a sapotrophic (SAP), ectomycorrhizal (EM), and arbuscular mycorrhizal (AM) fungus. 1 carbon acquisition by fungus, 2 transfer of fungal carbon to host plant, 3 remobilization of acquired plant carbon, and 4 plant excudates (including a Myc factor) that stimulates the fungal association.

In mixotrophs, light availability often mediates the energy derived from an autotrophic metabolism. In such cases, the niche of the mycoheterotrophic plant is more plastic than in obligate mycoheterotrophs, which are affected only indirectly by light availability (Preiss *et al.* 2011).

Key Microorganisms

It is important to note that no fungi is specially adapted to serve as a host for a mycoheterotrophic plant. It is quite the other way around--mycoheterotrophic plants have evolved strategies to extract carbon from (predominately) mycorrhizal hosts. These hosts include three primary functional groups: Ectomycorrhizas, Arbuscular mycorrhizas, and some Saprotrophs.

Most phylogenetic work of mycoheterotrophic fungi comes from the root systems of achlorophyllous orchdis (e.g. the genera *Corallorhiza, Epipogeum, Galeola, Gastrodia, Neottia,* and *Rhizanthella*) and the family Monotropoideae (e.g. the genera *Monotropa, Pterospora, and Sarcodes*) (Bidartondo 2005).

Ectomycorrhizas

Fungal hosts come primarily from the order Basidiomycota, with important families including Russulaceae, Tricholomataceae (esp. genus *Tricholoma*). Also common are fungi from the genus *Rhizopogon*, from the family Rhizopogonaceae and order Botales.

Arbuscular Mycorrhizas

Several mycoheterotrophic plant families associate with arbuscular mycorrhizal fungus. These include the Polygalaceae, Gentianaceae, Triuridaceae, Petrosaviaceae, Corsiaceae, and Burmanniaceae.

The fungal hosts confirmed of mycoheterotrophic plants includes species of the family Glomus group A, and at least one species of the genus *Gigaspora*.

Saprotrophs

The debate as to whether some mycoheterotrophs obtain carbon independently or from saprotrophic fungi (decomposers) is long-standing *Plants parasitic on fungi: unearthing the fungi in myco-heterotrophs and debunking the 'saprophytic' plant myth*. However, molecular tools have revealed that plants that obtain carbon from decomposing leaf litter due so through a saprotrphic fungal host.

For example, molecular methods have demonstrated that the achlorophyllous orchids *Epipogeum roseum* and *Fulophia zollingeri* associate with saprotrophic fungi from the family Coprinaceae (including *Corpirnus disseminatus, Psathyrella* spp., and *P. candolleana*) (Ogura-Tusjita and Yukawa 2008, Yamato *et al.* 2005). Acholorphyllous orchids are also known to associate with the fungal genus *Armillaria*, which is a known saprotroph and plant parasite.

Current Research

Although the seminal paper describing mycoheterotrophy dates back to 1888, surprisingly little is known about this fascinating plant-fungal interaction. This topic provides a review of some recent advances, as well as highlights reviews of tantalizing unknowns--and future research directions.

Mycoheterotrophic Generalist

The current paradigm in mycoheterotrophy celebrates the "unprecedented specificity" between plant and fungus. However, a recent study has demonstrated that some mycoheterotrophic plants are host-generalists, parasitizing multiple ectomycorrhizal taxa. Hynson and Bruns (2010) report that adults of ericaceous plant *Pyrola aphylla* can associate with as many as 15 fungal taxa. While previous studies have demonstrated that the mycoheterotroph seeds can germinate when exposed to innocula from multiple fungal species, this is the first study demonstrating that a plant can complete its life cycle without a specific fungal host.

Irradiance Mediated Mixotrophy

The shift between autotrophy and mycoheterophy in mixotrophic orchids may be fine tuned to light availability, even on short time scales. Preiss *et al.* 2011 exposed partially mycoheterotrophic species from the genus *Cephalanthera* to varying levels of irradiance. They find that the relative contribution of autotrophy and fungus is fine tuned to light availability, a finding that generally blurs the categories of partial and facultative mycoheterotroph.

Future Themes

A recent review issue of *New Phytologist* (2010) highlighted some important unknowns in the ecophysiology of mycoheterotrophic plant-fungal interactions.

(1) It is unknown whether the presence of mycoheterotrophic plants decreases the fitness of their host-fungus relative to unexploited, sympatric fungi.

(2) It is unknown how mycoheterotrophic plants recognize fungal hosts. The search for a general Myc factor, or signal molecule responsible for establishing the interaction and carbon and/or mineral nutrient exchange, continues. Have mycoheterotrophs co-opted signal molecules used to establish mutualistic mycorrhizal associations, or do they employ unique chemical signals?

(3) It is unknown what functional genes are associated with the mycoheterotrophic interaction, in either the plant or fungus.

References

- Peterson, R. Larry; Massicotte, Hugues B.; Melville, Lewis H. (2004). Mycorrhizas : anatomy and cell biology. Ottawa: NRC Research Press. ISBN 0851999018. OCLC 57587171

- Boswell, E. P.; R.T. Koide; D.L. Shumway; H.D. Addy. (1998). "Winter Wheat cover cropping, VA mycorrhizal fungi and maize growth and yield". Agriculture, Ecosystems and Environment. 67: 55–65. doi:10.1016/S0167-8809(97)00094-7

- "Six Years After Acquisition, Roche Quietly Shutters 454". Bio-IT World. 16 October 2013. Retrieved 13 March 2014

- Grant, C.; Bitman, S.; Montreal, M.; Plenchette, C.; Morel, C. (2005). "Soil and fertilizer phosphorus: effects on plant supply and mycorrhizal development". Canadian Journal of Plant Science. 85: 3–14. doi:10.4141/P03-182

- Smith, S. E.; Read, D. J.; Harley, J. L. (1997). Mycorrhizal symbiosis (2nd ed.). San Diego, Calif.: Academic Press. ISBN 0126528403. OCLC 35637899

- Martin, Francis. Molecular mycorrhizal symbiosis. Hoboken, New Jersey. ISBN 9781118951415. OCLC 958205579

- Sakamoto, Yuki; Yokoyama, Jun; Maki, Masayuki. "Mycorrhizal diversity of the orchid Cephalanthera longibracteata in Japan". Mycoscience. 56 (2): 183–189. doi:10.1016/j.myc.2014.06.002

- Alvarez, Marvin R. (1968). "Quantitative Changes in Nuclear DNA Accompanying Postgermination Embryonic Development in Vanda (Orchidaceae)". American Journal of Botany. 55 (9): 1036–1041. doi:10.2307/2440469. JSTOR 2440469

Understanding Fungiculture

Fungiculture refers to the cultivation of fungi for the production of food, medicine and other products. Many fungi are cultivated commercially across the globe, such as button mushroom, blewit, jelly fungi, winter mushroom, beech mushroom, shiitake, etc. This chapter has been carefully written to provide an in-depth understanding of fungiculture.

Fungiculture is the process of producing food, medicine, and other products by the cultivation of mushrooms and other fungi.

Fungiculture is the cultivation and production of edible and medicinal mushrooms. Mushrooms are the sporophores, or fruiting bodies, of filamentous fungi. Mushrooms are a nutritious food and an economically important commodity, extensively cultivated on a global scale. Mushroom production benefits from an understanding of a variety of habitat constraints and microbial interactions, upon which the success or failure of their cultivation depends. Although mushrooms themselves are macroscopic, their production is based on the manipulation of microbial habitat, community composition, or both, in the presence of the spores or mycelium of a desired fungal species, in order to create conditions favoring mycelial growth and mushroom formation. Along with fermentation and composting, fungiculture should be considered one of the first microbial biotechnologies.

All the cultivated mushrooms belong to the phylum Basidiomycota, and are all saprophytes. For the purposes of production, mushrooms can be roughly divided into two groups: primary and secondary decomposers. Though the boundary between the two groups is not absolute, the two groups of mushrooms require distinct cultivation techniques, and will be discussed separately. Some mushroom species are able to occupy different niches depending on environmental conditions (Stamets 2005).

Primary Decomposers

Niche

Primary decomposing fungi under cultivation include both wood-decay fungi, such as shitake (*Lentinula edodes*), oyster mushroom (*Pleurotus* spp.), and maitake (*Grifola frondosa*), and litter-decomposing fungi, such as winecap (*Stropharia rugosoannulata*).

The wood-decay fungi are divided, in turn, into two groups: brown-rot, which degrade cellulose and hemicellulose, and white-rot, which degrade lignin, as well as cellulose and hemicellulose. The white-rot fungi, in particular, play a crucial role in the global carbon cycle, by virtue of their ability to decompose large, complex lignin molecules, which constitute the most recalcitrant form of carbon found in plant material. Biodegradation of lignin is not thoroughly understood, but some of the more well-researched metabolic pathways involve the lignolytic enzymes manganese peroxidase, lignin peroxidase, and cellobiose dehydrogenase (Hattaka 1994).

It is the activity of these fungi that releases the nutrients and energy stored in the structural elements of plants, which get their strength and rigidity from an abundance of lignin, into a form usable by other organisms. Most of the wood-decaying fungi under cultivation are white-rot fungi, including the above-mentioned *L. edodes*, *Pleurotus* spp, and *G. frondosa*. Mushroom growers exploit the ability of fungi to digest substances that many organisms cannot, by pairing mushroom crops with semi-selective substrates that are nutritionally inaccessible to potential competitors (Stamets 2000).

Physical Environment

Substrates for the production of these mushrooms generally consist of dried, shredded, plant material, with very low nitrogen content. The C:N ratio in the wastes preferred by *L. edodes* and *P. ostreatus* range from 350:1 to 500:1, though the N content of these substrates is frequently supplemented with mineral fertilizer or high-N materials, such as rice bran (Chang & Miles, 2004). The most common material used for the production of wood-decaying fungi is sawdust, but techniques for growing nominally wood-decaying fungi on grasses are well-established for some mushroom crops. The shredding or pulverizing of the substrate material facilitates the ramification of the fungal mycelium throughout the substrate. This is significant for mushroom producers for several reasons. The speed of colonization, together with other factors, controls the time necessary to produce a crop, and the amount of substrate accessed by the mycelium controls the yield per unit. A fine, granular substrate structure, which permits faster and easier ramification, consolidation, and breakdown, by fungal mycelia, will produce higher yields, sooner, than bulkier substrates.

Microbial Processes

Competition-Contamination

The preference of primary decomposer wood-decay fungi for the relatively microbially simple environment found within undecayed wood, means that these fungi can be very vulnerable to microbial competition, when present. This makes manipulation of the microbial environment fairly straight-forward for mushroom growers - in theory. Most commercial cultivation involves sterilization of the substrate, prior to inoculation with the desired species.

In practice, fungi growing in sterilized substrates are vulnerable to contamination by other fungal or bacterial organisms. The oligomers resulting from the breakdown of lignin and cellulose by fungi are a viable source of nutrition for a wide range of bacteria as well - and fungi grown in sterile environments are less resilient to the presence of bacteria than they are in their natural habitat. In the wild, fungal-bacterial interactions range from competitive to mutualistic (de Boer 2008). In contrast, in the cultivation of primary decomposers, the presence of any other fungal or bacterial organism in a given container of mushroom substrate is usually regarded as a total loss of that container's production.

Secondary Decomposers

Niche

Secondary decomposer fungi rely on the metabolic products and by-products of fungal, bacterial, and other primary decomposers. Living in soil or rotting wood, they prefer environments that are more biologically complex than the primary decomposers, and are cultivated on specialized composts. Commonly cultivated secondary-decomposer fungi include the common button mushroom (*Agaricus bisporus*), largely in temperate areas, and the paddy straw mushroom (*Volvariella volvaceae*), in the tropics.

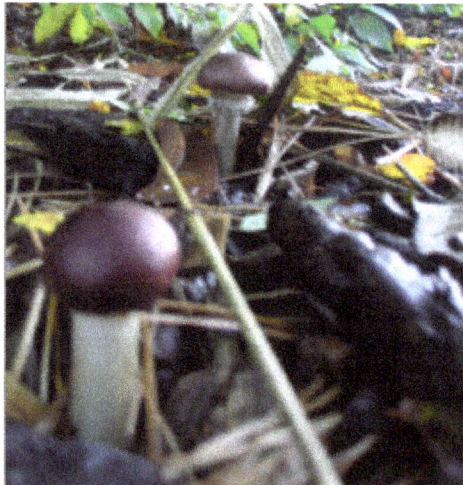

Stropharia rugosoannulata in its natural habitat in the forest understory

The niches of primary and secondary decomposition are not perfectly discrete. Despite their ecological role as primary decomposers, oyster mushrooms can be cultivated in the style of secondary decomposers, on composted substrates (Vajna 2010). The litter-decomposing fungi naturally occupy a niche that combines elements of both primary and secondary decomposition. S. rugosoannulata is a primary decomposer, and can digest a variety of fresh coarse lignocellulosic debris. But as a litter decomposer, it occupies zones of high biological complexity on the forest floor, and thrives in the presence of bacteria (Stamets 2005). This resilience creates the opportunity for a naturalistic style of cultivation of this mushroom that is difficult to achieve with other mushrooms.

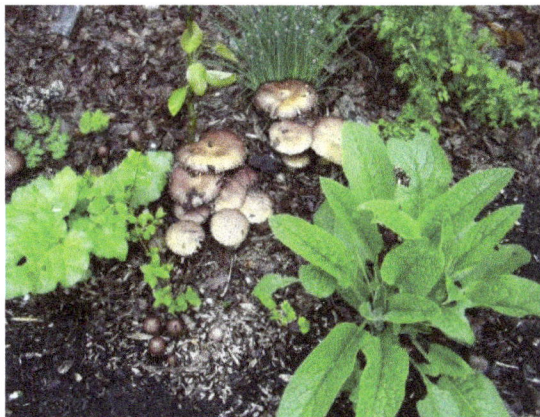

Stropharia rugosoannulata grown in a naturalistic outdoor polyculture with perennial and annual vegetables

Physical Environment

Secondary decomposers are grown in composted substrates that have already undergone a significant amount of decomposition. Mushroom composts are produced with specialized rapid composting techniques, utilizing straw and manures (for *A. bisporus*) and many other combinations of agricultural residues and plant wastes for other mushroom species. These composts have a much higher proportion of nitrogen than the substrates preferred by the primary decomposers. Optima for C:N range from 17:1, in the case of the common white button mushroom, *Agaricus bisporus*, to 80:1, for the paddy straw mushroom, *V. volvacea* (Chang & Miles, 2004).

Biological Interactions

Succession

Unlike the primary decomposers, cultivating secondary decomposer fungi involves a sophisticated manipulation (rather than simple suppression) of microbial activity. Like other composts, mushroom composts undergo a successional process, as the activity of bacteria and non-cropping fungi alter the eco-physiological parameters of the compost pile environment, especially temperature, available nutrients, and pH. These eco-physiological parameters trigger changes in community composition. The mushroom compost production process, however, is unusually tightly controlled, and therefore quite rapid - usually taking place over three weeks. Like other composts, mushroom composts contain a wide spectrum of microbial diversity, passing through initial mesophile-dominanted communities, to thermophilic, and then back to (new) mesophilic commmunities in the mature compost. *Thermoactinomyces* species of thermophilic actinomycetes, which along with *Thermobifida fusca* are associated with occupational respiratory illness, have been shown to be common in mushroom composts (Song et al. 2001). Contrary to expectations about the metabolic activity in mature composts, finished mushroom composts have demonstrated the abundance of a consortium of supposedly cellulose-degrading bacteria, related to *Pseudoxanthomonas*, *Thermobifida*, and *Thermomonospora* (Székely et al., 2009).

Large-scale mechanized windrow composting for *A. bisporus* production

Predation

Secondary decomposers prefer much more biologically complex environments. They not only tolerate the presence of bacteria, but have been shown to draw on both dead and living bacterial biomass as a nutrient source (Barron 1988, Fermor 1991). This ability to degrade living bacterial biomass has generated an interest in the use of mushroom mycelium to filter agricultural run-off, in order to reduce the presence of *E. coli* and coliform bacteria, an emerging technique dubbed "mycofiltration" (Stamets 2005).

Competition

Adaptiveness to biologically complex environments notwithstanding, secondary decomposers are still vulnerable to microbial competition. *Trichoderma* spp., particularly *T. harzianum* and *T. aggressivum*, are the most notorious fungal competitors in production of Agaricus and other mushrooms (Savoie & Mata, 2003). The bacterial pathogen Pseudomonas tolaasii causes a condition called "bacterial blotch," resulting in discoloration of mushroom caps, and reduction or loss of production value. (Stamets & Chilton, 1983).

Mushroom

Mushroom, the conspicuous umbrella-shaped fruiting body (sporophore) of certain fungi, typically of the order Agaricales in the phylum Basidiomycota but also of some other groups. Popularly, the term mushroom is used to identify the edible sporophores; the term toadstool is often reserved for inedible or poisonous sporophores. There is, however, no scientific distinction between the two names, and either can be properly applied to any fleshy fungus fruiting structure. In a very restricted sense, mushroom indicates the common edible fungus of fields and meadows (Agaricus campestris). A very closely related species, A. bisporus, is the mushroom grown commercially and seen in markets.

Umbrella-shaped sporophores are found chiefly in the agaric family (Agaricaceae), members of which bear thin, bladelike gills on the undersurface of the cap from which the spores are shed. The sporophore of an agaric consists of a cap (pileus) and a stalk (stipe). The sporophore emerges from an extensive underground network of threadlike strands (mycelium). An example of an agaric is the honey mushroom (*Armillaria mellea*). Mushroom mycelia may live hundreds of years or die in a few months, depending on the available foodsupply. As long as nourishment is available and temperature and moisture are suitable, a mycelium will produce a new crop of sporophores each year during its fruiting season.

Fruiting bodies of some mushrooms occur in arcs or rings called fairy rings. The mycelium starts from a spore falling in a favourable spot and producing strands (hyphae) that grow out in all directions, eventually forming a circular mat of underground hyphal threads. Fruiting bodies, produced near the edge of this mat, may widen the ring for hundreds of years.

A few mushrooms belong to the order Boletales, which bear pores in an easily detachable layer on the underside of the cap. The agarics and boletes include most of the forms known as mushrooms. Other groups of fungi, however, are considered to be mushrooms, at least by laymen. Among these are the hydnums or hedgehog mushrooms, which have teeth, spines, or warts on the undersurface of the cap (e.g., *Dentinum repandum*, *Hydnum imbricatum*) or at the ends of branches (e.g., *H. coralloides*, *Hericium caput-ursi*). The polypores, shelf fungi, or bracket fungi (order Polyporales) have tubes under the cap as in the boletes, but they are not in an easily separable layer. Polypores usually grow on living or dead trees, sometimes as destructive pests. Many of them renew growth each year and thus produce annual growth layers by which their age can be estimated. Examples include the dryad's saddle (*Polyporus squamosus*), the beefsteak fungus (*Fistulina hepatica*), the sulfur fungus (*P. sulphureus*), the artist's fungus (*Ganoderma applanatum*, or *Fomes applanatus*), and species of the genus *Trametes*. The clavarias, or club fungi (e.g., *Clavaria*, *Ramaria*), are shrublike, clublike, or coral-like in growth habit. One club fungus, the cauliflower fungus (*Sparassis crispa*), has flattened clustered branches that lie close together, giving the appearance of the vegetable cauliflower. The cantharelloid fungi (*Cantharellus* and its relatives) are club-, cone-, or trumpet-shaped mushroomlike forms with an expanded top bearing coarsely folded ridges along the underside and descending along the stalk. Examples include the highly prized edible chanterelle (*C. cibarius*) and the horn-of-plenty mushroom (*Craterellus cornucopioides*). Puffballs (family Lycoperdaceae), stinkhorns, earthstars (a kind of puffball), and bird's nest fungi are usually treated with the mushrooms. The morels (*Morchella*, *Verpa*) and false morels or lorchels (*Gyromitra*, *Helvella*) of the phylum Ascomycota are popularly included with the true mushrooms because of their shape and fleshy structure; they resemble a deeply folded or pitted conelike sponge at the top of a hollow stem. Some are among the most highly prized edible fungi (e.g., *Morchella esculenta*). Another group of ascomycetes includes the cup fungi, with a cuplike or dishlike fruiting structure, sometimes highly coloured.

Other unusual forms, not closely related to the true mushrooms but often included with them, are the jelly fungi (*Tremella* species), the ear fungus or Jew's ear (*Auricularia auriculara-judae*), and the edible truffle.

Edible Mushroom

Edibility is defined as the absence of poisons and the presence of a desirable taste and smell.

There are many edible mushrooms that can be found that are good to eat that are very distinct and will not be mistaken for poisonous species. If the collector sticks to those species, there should not be any problems. The most common reason for the occurrence of mushroom poisoning is collecting and eating a species of mushroom that closely resembles one that is poisonous. This is probably the most common cause of mushroom poisoning.

A question that is often asked by the novice mushroom collector is how can you determine if a mushroom is poisonous or not. Often the poisonous species is referred to as a "toadstool", but in terms of the general appearance of poisonous mushrooms do not look any different than those that are edible. The toadstool label is just a means by which some people refer to poisonous mushrooms. There are no generalizations that can be made that will allow you to distinguish between what is edible and what is not. "Fool-proof" means of making this distinction are myths and are untrue. Some of the most common example is that a silver coin will turn black when cooked with a poisonous mushroom, if animals are seen eating the mushroom, then it is not poisonous, if the top layer of the cap can be peeled, it is not poisonous and there are more. The only means by which species have been determined to be poisonous is if somebody had eaten it and got sick and/or died

There are all sorts of different mushrooms in the world.

From common edible mushrooms to exotic varieties from the far-east, the earth offers thousands of them.

Although they are technically a type of fungus, mushrooms are commonly recognized as a vegetable.

No matter what dietary system you prefer, whether it's a keto diet or vegan, mushrooms can play a significant role in health.

So, here are nine of the most delicious types of mushrooms you can eat.

1. Cremini Mushrooms

Cremini mushrooms are a variety of fungus that belongs to the white button mushroom family.

The species name is Agaricus Bisporus, and this family of mushroom also includes portobello.

These three mushrooms—cremini, portobello and white button—are the three most commonly consumed in the world (1).

That being said, these three mushrooms are actually the same mushroom. Although they look different, the varying appearance just depends on age.

White button mushrooms are the freshest and youngest, then cremini, and portobellos have been left to mature for a long time.

Compared to white buttons, cremini mushrooms have a browner color, a meatier texture, and a deeper flavor.

Nutrition wise, cremini mushrooms provide an excellent source of the following micronutrients (2);

- Selenium: 37% RDA

- Riboflavin (Vitamin B2): 29% RDA

- Copper: 25% RDA

- Niacin: 19% RDA

- Pantothenic Acid (Vitamin B5): 15% RDA

- Potassium: 13% RDA

- Phosphorus: 12% RDA

2. Morel Mushrooms

These are certainly one of the most unusual types of edible mushrooms.

Morel mushrooms *(Morchella esculenta)* look like honeycomb on a stick, and they have a strange, mysterious appearance which suggests we probably shouldn't eat them.

However, eat them we can, and they taste as unique as they look.

These are like that, only much more intense and they have a kind of nutty flavor too.

In short, if you like shiitake you'll probably love Morels.

Morels are Wild—Not Commercial—Mushrooms

It's hard to cultivate morel mushrooms on a large scale, so it's rare to find them in a store.

However, we can pick our own, or we can grow them ourselves.

Due to their unique appearance, many people acknowledge that morel mushrooms are the easiest—and safest—wild mushroom to identify

In regard to nutrition, morel mushrooms offer significant amounts of;

- Iron: 68% RDA

- Vitamin D: 52% RDA

- Copper: 31% RDA

- Manganese: 29% RDA

- Phosphorus: 19% RDA

- Zinc: 14% RDA

- Riboflavin (Vitamin B2): 12% RDA

3. Shiitake Mushrooms

Shiitake (*Lentinula* edodes) is one of the most infamous types of mushroom and for a good reason; they taste delicious.

With a meaty, chewy taste, they go well with almost everything.

Enjoying the most popularity in Japan—their homeland—they are famous for having a variety of health benefits.

For instance, they are particularly renowned for their anticarcinogenic and antimicrobial properties.

A wealth of studies show that they may help protect against several medical conditions;

- In a randomized dietary study of 52 healthy men and women, those given dried shiitake mushrooms showed improved immunity and lower levels of CRP (a marker of inflammation). On the positive side, the dosage was at realistic consumption levels in everyday life.

Inflammation is a leading cause of most chronic disease, and immunity helps our body fight illness, so these are excellent benefits.

Per 100g, shiitake mushrooms provide the following vitamins and minerals in large amounts;

- Copper: 45% RDA

- Pantothenic Acid: 36% RDA

- Riboflavin: 10% RDA

- Manganese: 10% RDA

- Zinc: 9% RDA

- Vitamin B6: 8% RDA

- Niacin: 7% RDA

4. Oyster Mushrooms

As shown above, king oyster mushrooms are one of the biggest types of edible mushrooms.

They have a striking appearance and are very thick in shape, giving them an extremely chewy and spongy texture — a little bit like squid.

For this reason, roasting king oysters in the oven works best and leaves you with a deep, rich flavor — especially if you add a bit of butter and salt beforehand.

The mushroom goes by the scientific name of Pleurotus eryngii and it is native to Europe, the Middle East, and North Africa.

The mushroom is easy to identify in the wild and, outside of Japan and Australia, has no poisonous look-alikes.

Nutrient-wise, king oysters provide the following micronutrients per 100g;

- Niacin: 25% RDA

- Riboflavin: 21% RDA

- Pantothenic Acid: 13% RDA

- Copper: 12% RDA

- Phosphorus: 12% RDA

- Potassium: 12% RDA

- Iron: 7% RDA

Health Benefits

Oyster mushrooms are a species which have long been used as a medicinal mushroom.

Notably, studies show that they contain a significant amount of antioxidants and that their compounds exert anti-tumor and anti-inflammatory effects.

Also, consumption of this edible mushroom appears to lower triglycerides and the LDL/HDL ratio.

Although they can be a little expensive, you can pick them up for a lower price if you go to a Chinese/Asian market.

5. Lion's Mane Mushrooms

Similar to morel mushrooms, the lion's mane mushroom (Hericium erinaceus) has an unusual appearance.

It is a species of mushroom...that doesn't look like one.

It is large, edible and certainly an interesting looking fungus, as you can see in this picture;

Significantly, research indicates that lion's mane mushrooms have anti-inflammatory, gastroprotective, cardioprotective, and an inhibitory effect on cancer metastasis.

These mushrooms grow in the wild throughout Europe, North America and Asia. Although relatively rare in Western dishes, it plays a large part in Chinese cuisine.

Despite this, the mushroom has become popular in the health and supplement industry, and a variety of products are available. These include extracts, tablets, and even coffee-mix drinks.

However, rather than buying extracts, you can buy the real thing if you hunt around in some Asian grocery stores.

Otherwise, you can also buy them as dried mushrooms.

They taste pretty good and have a very intense meaty flavor.

6. Enoki Mushrooms

Enoki mushrooms (*enokitake*) are long thin white mushrooms which resemble a piece of string.

Again, they are one of the more unique looking mushroom varieties.

Enoki mushrooms taste great; they are also relatively simple to grow and cheap to buy.

In certain Asian and Italian dishes, they can act as a replacement for noodles and spaghetti due to their chewy texture and noodle-like appearance.

For the same reason, tossing some into a stir-fry has great results.

Here are their most significant nutrients on a per-100g basis;

- Niacin: 30% RDA
- Folate: 13% RDA
- Thiamin: 12% RDA
- Potassium: 11% RDA
- Pantothenic Acid: 11% RDA
- Phosphorus: 11% RDA
- Riboflavin: 10% RDA

7. Button Mushrooms

Button mushrooms (*Agaricus bisporus)* are the baby version of shiitake and cremini; they are still very fresh and at an early-life stage.

These white mushrooms are probably the most common—and widespread—variety in the world. In fact, they represent 90% of the edible mushrooms consumed in the United States.

Despite a common belief that these mushrooms aren't as healthy as other types, they have some impressive health benefits.

First of all, their nutrient profile. Button mushrooms contain the following vitamins and minerals in substantial amounts;

- Riboflavin: 24% RDA

- Niacin: 18% RDA

- Copper: 16% RDA

- Pantothenic Acid: 15% RDA

- Selenium: 13% RDA

- Potassium: 9% RDA

- Phosphorus: 9% RDA

Although they are not usually mentioned alongside the "medicinal mushroom" tag, button mushrooms have shown promise in clinical studies.

In particular, here are a couple of recent research findings;

- White button mushrooms enhance the strength of cells critical to the body's immune system.

- In a study involving 24 healthy volunteers, 12 were assigned to eat a diet that included 100g button mushrooms daily, and the remaining 12 ate the same control diet except for the button mushrooms. Over two weeks, secretory immunoglobulin—an antibody involved in the immune system—increased by 56% in the button mushroom group only.

8. Portobello Mushrooms

If button mushrooms are the babies, then portobello mushrooms are the grandparent!

Portobello mushrooms are the same species—*Agaricus bisporus*—but at a late stage of life.

The mushrooms are therefore much bigger and wider in diameter, as well as being deeper and richer in flavor.

For me, they are one of the best types of mushrooms and baked portobellos are delicious — especially when they are stuffed with some cheese.

The mushroom provides the following major nutrients;

- Riboflavin: 28% RDA

- Niacin: 23% RDA

- Copper: 20% RDA

- Niacin: 23% RDA

- Selenium: 16% RDA

- Pantothenic Acid: 15% RDA

- Potassium: 14% RDA

9. Porcini Mushrooms

Porcini mushrooms (*Boletus edulis*) are one of the most popular sorts of mushrooms for culinary purposes.

Taste wise, they have a deep and mild nutty flavor with an intense aroma. Porcinis can be either purchased fresh at markets or in dried form.

Porcini mushrooms are also an attractive target for wild mushroom foragers, due to their easily identifiable features.

Animal studies suggest that extracts of these mushrooms help protect against cardio-vascular disease.

For instance, repeated daily doses of the mushroom resulted in significant decreases in blood pressure, triglycerides, and an increase in HDL.

Medicinal Mushroom and Fungi

Fungi are more closely related to us than they are to plants. This means their physiology has distinct therapeutic potential, since many of the molecules they produce are evolutionarily compatible with our own bodies. Mushrooms, the fruiting body of fungi, have been widely popular across many traditions for thousands of years. In addition to edible mushrooms, people living in the Americas and Asia (particularly China) used dozens of mushroom species for therapeutic purposes. While largely absent from ancient Greek traditions, medicinal mushrooms are becoming more popular in Western traditions as clinical trials repeatedly demonstrate tumor suppression, chemotherapy/radiation support, and immune modulating effects.

Mushrooms are particularly powerful immune modulators. They're composed of complex polysaccharides, long-chain sugars, called beta-glucans. These molecules are robust enough to make it through our acidic stomach and into the small intestine where they interact with the diverse landscape in the mucus membrane lining of our gut. This is the same mechanism through which Echinacea and Astragalus act to improve our immune response.

Turkey Tail Mushroom

First, these complex sugars provide our microflora with a nourishing meal. In turn, our bacteria help to break down these polysaccharides into oligosaccharides, which then act as signaling molecules in our gut-associated lymph tissue. Through a variety of signaling cascades, these oligosaccharides can stimulate the production of immune cells that can migrate to other tissues, such as bone marrow where new lymph cells are created. Alternatively, some types of polysaccharides can suppress chronic inflammatory immune condition (known as "latent heat" in Traditional Chinese Medicine) or overgrowth of a variety of cancerous cell lines.

Besides the occasional allergic reaction, there are no known harmful effects of mushrooms, indicating their overall effectiveness in restoring balance to an immune disregulation. In addition to polysaccharides, medicinal mushrooms contain triterpenes (terpenes) with more specialized effects such as antiviral, antioxidant, hypotensive, hypocholesterolemic, hepatoprotectant, and antifibrotic constituents.

Health Benefits of Medicinal Mushrooms

While much attention in recent years has focused on various immunological and anti-cancer properties of certain mushrooms, they also offer other potentially important health benefits, including antioxidants, anti-hypertensive and cholesterol-lowering properties, liver protection, as well as anti-inflammatory, anti-diabetic, anti-viral and anti-microbial properties. These properties have attracted the interest of many pharmaceutical companies, which are viewing the medicinal mushroom as a rich source of innovative biomedical molecules.

Below is a brief description of 5 popular medicinal mushroom varieties: Coriolus versicolor (Turkey Tails), Ganoderma lucidum (Reishi), Agaricus blazei (add the common name Himematsutake), Polyporus umbellatus (add the common name Zhu Ling), Hericium erinaceus (Lion's Mane).

Coriolus Versicolor (Turkey Tails)

Coriolus versicolor ("multicolored mushroom"), also known as Trametes versicolor, is a mushroom readily found in woodlands in China and Europe and is the most commonly found polypore in the oak woods of the Pacific Coast in the U.S. It grows in clusters or tiers on fallen hardwood trees and branches, frequently in large colonies. As its name implies, it is often multi-colored, with contrasting concentric bands, variously appearing in shades of white, gray, brown, black, blue or even red. It has a thin, velvety fruiting body, usually 2- 7 cm wide, fans out into wavy rosettes, giving rise to its popular name, Turkey Tails.

Uses: Anti-cancer action: PSK has been shown to be effective against several cancers, including cervical cancer, in combination with other therapeutic agents; appears to enhance the effect of radiation therapy; PSP significantly lessened the side effects of conventional medical protocols used in the treatment of cancers of the esophagus, stomach and lungs, as well as significantly increasing the rate of remission in esophageal cancers. Cardiovascular health: Lowered cholesterol in animal studies. Immune enhancement: PSK increases interferon production, as well as scavenging superoxide and hydroxyl free radicals, has demonstrated anti-viral activity, possibly even inhibiting HIV infection.

Ganoderma Lucidum (Reishi)

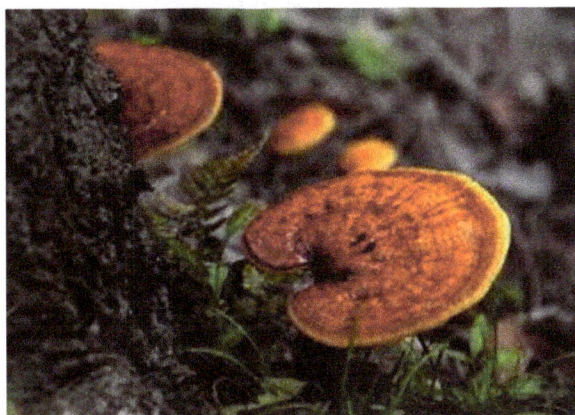

Ganoderma lucidum ("shining skin") is a visually striking polypore with a hard woody texture and a shiny, varnished appearance. It primarily grows on oaks, plum trees and other hardwoods, and has a 2-20 cm semi-circular or kidney-shaped cap, variously colored white, yellow, blue, red, purple or black. Ganoderma species are found worldwide, though the Chinese and Japanese species have been studied the most extensively for their therapeutic value. It is somewhat rare in the wild, and so in recent years has been commercially cultivated, making it more widely available. In the West it is usually known by its Japanese name, reishi.

Uses: Athletic performance: Enhances oxygenation of the blood, reducing and preventing altitude sickness in high altitude mountain climbers. Cardiovascular health: Lowers cholesterol levels, reduced blood and plasma viscosity in a controlled study of patients with high blood pressure and high cholesterol. Immune enhancement: Potent action against sarcoma, stimulates macrophages and increases levels of tumor-necrosis factor (TNF-α) and interleukins. Immunopotentiation: Anti-HIV in in vitro and in vivo animal studies; protects against ionizing radiation. Liver health: Reduced liver enzyme levels (SGOT and SGPT) in hepatitis B patients. Respiratory health: 60-90 % of 3,000 patients with chronic bronchitis showed clinical improvement, especially older patients with bronchial asthma. Regenerates bronchial epithelium (bronchial tract lining).

Polyporus Umbellatus (Zhu Ling)

Polyporus umbellatus ("umbrella-like polypore"), also known as Grifola umbellata, is a white-to-gray mushroom that grows in dense rosettes from a single stem. It is found in deciduous woodlands in China, Europe and Eastern and Central North America, growing from dead tree stumps or the roots of birches, maples, beeches and willows.

Uses: Anti-cancer actions: Used in the treatment of lung and other cancers; has demonstrated pronounced anti-tumor activity in in vitro and in vivo animal studies; helps reduce the side-effects of chemotherapy. Immune enhancement: stimulates and enhances the performance of the immune system and accelerates production of IgM and strengthens the power of monocytes. Liver health: can help alleviate symptoms of chronic hepatitis; was used as part of an herbal formula that cured 17 of 39 patients with cirrhosis of the liver, and brought about significant improvement in 19 others.

Hericium Erinaceus (Lion's Mane)

Hericium erinaceus ("spiny hedgehog") is a snow-white, globe-shaped fungus composed of downward cascading, icicle-shaped spines. Its striking appearance gives rise to its various common names, Lion's Mane, Monkey's Head and Hedgehog Fungus. It grows up to 40 cm in diameter on dead or dying broadleaf trees — such as oak, walnut, maple and sycamore — and is found in China and Japan, as well as parts of Europe and North America. It is considered a gourmet mushroom, long popular with forest folk, with a flavor variously described as reminiscent of lobster or eggplant. Uses: Anti-cancer effects: Helps in the treatment of esophageal and gastric cancers, may extend the life-span of cancer patients. Digestive enhancement: Promotes proper digestion; effective against gastric and duodenal ulcers and gastritis. Immune enhancement: Protects the gastrointestinal tract against environmental toxins, inflammation and tumor formation, an extract was used as part of a protocol that helped increase T and B lymphocytes in mice.

Penicillium

Penicillium is a group (Genus) of moulds found everywhere world-wide. It is the mould that saved millions of lives by producing the first ever known modern antibiotic, the penicillin. The discovery of penicillin from the fungus *Penicillium chrysogenum* (then known as *Penicillium notatum*) by Sir Alexander Fleming in 1928, perfected the treatment of bacterial infections.

Penicillium chrysogenum spores

The name *Penicillium* comes from the resemblance of the spore producing structures (conidiophores) of the fungus to a paintbrush (penicillus is the Latin word for paintbrush). They are found in soil, decaying vegetation, air and are common contaminants on various substances.

Penicillium causes food spoilage, colonizes leather objects and is an indicator organism for dampness indoors. Some species are known to produce toxic compounds (mycotoxins). The spores can trigger allergic reactions in individuals sensitive to mould. Therefore, the health of occupants may be adversely affected in an environment that has an amplification of *Penicillium*.

About 200 species of *Penicillium* have been described. They are commonly called the blue or green moulds because they produce enormous quantities of greenish, bluish or yellowish spores which give them their characteristic colours. Spores from this species of mold are found everywhere in the air and soil. As mentioned earlier, *Penicillium* species are one of the most common causes of spoilage of fruits and vegetables. For example, *P. italicum* and *P. digitatum* are common causes of rot of citrus fruits, while *P. expansum* is known to spoil apples.

Penicillium chrysogenum is the most common species in indoor environment. It is widespread and has a wide range of habitats. In indoor environment, it is extremely common on damp building materials, walls and wallpaper, floor, carpet mattress and upholstered furniture dust. It produces a number of toxins of moderate toxicity. It is allergenic (i.e., it can trigger allergic reactions).

Some species of *Penicillium* can also infect immunocompromised individuals. For example, *P. marneffei* is pathogenic particularly in patients with AIDS and its isolation from blood is considered as an HIV marker in endemic areas. It has emerged as the third most common opportunistic pathogen among HIV-positive individuals in Southeast Asia where it is endemic and infects bamboo rats which serve as reservoirs for human infections.

Penicillium as A Producer of Mycotoxins

Penicillium species other than *P. marneffei* are commonly considered as contaminants but they are also known to produce mycotoxins. For example, *P. verrucosum* produces a mycotoxin, ochratoxin A , which is damaging to the kidney (nephrotoxic) and could be cancer causing (carcinogenic). The production of the toxin usually occurs in cereal grains at cold climates but has been isolated in buildings contaminated with *Penicillium*. Other mycotoxins include patulin, citrinin, and citroviridin among other

Blue Cheese

Blue Vein cheeses also called Blue cheese is a generic term used to describe cheese produced with cow's milk, sheep's milk, or goat's milk and ripened with cultures of the mold Penicillium. The final product is characterized by green, grey, blue or black veins or spots of mold throughout the body. These veins are created during the production stage when cheese is 'spiked' with stainless steel rods to let oxygen circulate and encourage the growth of the mold. This process also softens the texture and develops the distinctive blue flavour

Blue moulds have a particularly unique effect on cheese. They accelerate two processes dramatically: proteolysis (breakdown of proteins), which causes the cheese to take on an extra-creamy texture (especially in proximity to the blue mould veins) and lipolysis (breakdown of fats), which makes up the tangy, spicy, sharp and strong flavour. The creamy texture stands up to the sharp flavour and together they produce an exciting flavour/texture/aroma profile, which is often further balanced against sweet/nutty milk and lots of salt (blue cheeses typically contain twice the salt of other cheeses). This combination is so unique – it is unlike any other food.

The origin of Blue cheese has an interesting story. It is thought to have been invented by accident when a drunken cheese maker left behind a half-eaten loaf of bread in moist cheese caves. When he returned back, he discovered that the mold covering the bread had transformed it into a blue cheese.

Blue cheese is also identified by a peculiar smell that comes from the cultivated bacteria. The flavour of the cheese depends on the type of blue cheese, shape, size, climate of the curing and the length of ageing. But it generally tends to be sharp and salty. Some of the famous blue cheeses around the world are Roquefort from France, Gorgonzola from Italy and Stilton from England.

Blue cheese tastes best when served with crackers, pears, raisins, fruit breads and walnuts. Crumble the cheese and melt it into sour cream, plain yogurt or mayonnaise as a dressing.

Penicillium Roqueforti

Penicillium fungi derive their name from the Latin word for paint brush, which is due to the shape of their spore forming organ called the conidiophore. When examined under a microscope, it resembles a paint brush.

Useful Fungus

Penicillium fungi are known as the source of antibiotics - a convenient property accidentally discovered by microbiologist Alexander Fleming back in in 1928. Penicillium roqueforti also produces a number of antibiotics, but is most famous for its use in the food industry.

Blue Cheese

In fact, P. roqueforti is absolutely essential in the making of a number of different cheeses, such as Roquefort cheese (hence its name). During the production process, the cheese is punctured by needles, so that oxygen can get all the way to the core of the cheese. P. roqueforti, which is added to the cheese at an early stage, needs this oxygen in order to grow and give the cheese its characteristic flavour, smell and colour. The blue colour is a result of the fungus's spores. Blue cheese, likely containing P. roqueforti, was first described as early as AD 79.

Food Gone Bad

Penicillium roqueforti is a type of fungus that likes to grow on food, and not always in a manner that we can appreciate. If you leave a piece of bread out for too long, then P. roqueforti will have its way with it. It will create a network of fungal threads , or hyphae, all over the bread as well as inside it, and cause an unpleasant odour.

Permissions

We would like to thank the editorial team for lending their expertise to make the book truly unique. They have played a crucial role in the development of this book. Without their invaluable contributions this book wouldn't have been possible. They have made vital efforts to compile up to date information on the varied aspects of this subject to make this book a valuable addition to the collection of many professionals and students.

This book was conceptualized with the vision of imparting up-to-date and integrated information in this field. To ensure the same, a matchless editorial board was set up. Every individual on the board went through rigorous rounds of assessment to prove their worth. After which they invested a large part of their time researching and compiling the most relevant data for our readers.

The editorial board has been involved in producing this book since its inception. They have spent rigorous hours researching and exploring the diverse topics which have resulted in the successful publishing of this book. They have passed on their knowledge of decades through this book. To expedite this challenging task, the publisher supported the team at every step. A small team of assistant editors was also appointed to further simplify the editing procedure and attain best results for the readers.

Apart from the editorial board, the designing team has also invested a significant amount of their time in understanding the subject and creating the most relevant covers. They scrutinized every image to scout for the most suitable representation of the subject and create an appropriate cover for the book.

The publishing team has been an ardent support to the editorial, designing and production team. Their endless efforts to recruit the best for this project, has resulted in the accomplishment of this book. They are a veteran in the field of academics and their pool of knowledge is as vast as their experience in printing. Their expertise and guidance has proved useful at every step. Their uncompromising quality standards have made this book an exceptional effort. Their encouragement from time to time has been an inspiration for everyone.

The publisher and the editorial board hope that this book will prove to be a valuable piece of knowledge for students, practitioners and scholars across the globe.

Index

www.ingramcontent.com/pod-product-compliance
Lightning Source LLC
Chambersburg PA
CBHW062003190326
41458CB00009B/2956